T0145363

Advanced Information and Knowledge Processing

Information systems and intelligent knowledge processing are playing an increasing role in business, science and technology. Recently, advanced information systems have evolved to facilitate the co-evolution of human and information networks within communities. These advanced information systems use various paradigms including artificial intelligence, knowledge management, and neural science as well as conventional information processing paradigms.

The aim of this series is to publish books on new designs and applications of advanced information and knowledge processing paradigms in areas including but not limited to aviation, business, security, education, engineering, health, management, and science.

Books in the series should have a strong focus on information processing - preferably combined with, or extended by, new results from adjacent sciences. Proposals for research monographs, reference books, coherently integrated multi-author edited books, and handbooks will be considered for the series and each proposal will be reviewed by the Series Editors, with additional reviews from the editorial board and independent reviewers where appropriate. Titles published within the Advanced Information and Knowledge Processing Series are included in Thomson Reuters' Book Citation Index and Scopus.

More information about this series at https://link.springer.com/bookseries/4738

Israël César Lerman · Henri Leredde

Seriation in Combinatorial and Statistical Data Analysis

 Springer

Israël César Lerman
Data and Knowledge Management
Department
University of Rennes 1, IRISA
Rennes, Ille-et-Vilaine, France

Henri Leredde
Laboratoire Analyse Géométrie et
Applications (LAGA, CNRS UMR 7539)
Sorbonne Paris-Nord University
Villetaneuse, Seine-Saint-Denis, France

ISSN 1610-3947   ISSN 2197-8441   (electronic)
Advanced Information and Knowledge Processing
ISBN 978-3-030-92696-0   ISBN 978-3-030-92694-6   (eBook)
https://doi.org/10.1007/978-3-030-92694-6

This Springer imprint is published by the registered company Springer Nature Switzerland AG
The registered company address is: Gewerbestrasse 11, 6330 Cham, Switzerland

Preface

To begin, let us describe the circumstances and motivations that led to the publication of this book. This publication can be related to that of the book entiteled

Foundations and Methods in Combinatorial and Statistical
    Data Analysis and Clustering

published in the same Springer Nature series in 2016. Let us recall that the latter book was a new recasting and comprehensive english version of an important part of the french book:

Classification et Analyse Ordinale des Données

published with the support of the CNRS—by Dunod (Paris) in 1981.

Clearly, in this version, the immense development of the domain has been taken into account. It was Dan A. Simovici, Professor at the University of Massachusetts (Department of Computer Science) who encouraged me to propose this publication. I am very grateful to him.

In the development of the English book above, denoting by E the set to be statistically structured, we are led to provide E with a *symmetrical* numerical or ordinal similarity or dissimilarity over E. This symmetry is with respect to $E \times E$. On the other hand and above all, the synthetic structure built on E is *symmetrical*. The latter structure is defined by either a classification tree (hierarchical clustering) or by a classification (non-hierarchical clustering) of E.

In combinatorial data analysis, the synthetic structure may be ordinal. In this case, relative to a given ordered pair (x, y) in $E \times E$, x is to y what is not y to x.

The initial idea of a second volume was to concentrate the matter on methods for which the synthetic structure built on E is *asymmetrical*. Mostly, this relation has an ordinal nature. It might be discrete (logical) or valued, numerically. Now, regarding the initial similarity (resp., dissimilarity) measure given on E, the latter might be symmetrical or oriented. The subjects of ordinal data analysis which are taken into account in the french book (Dunod 1981) are:

- Principal Component Analysis and Correspondence Analysis;
- Formal and Methodological Comparison Between Component Factorial Approach and Classification Approach;
- Combinatorial and Statistical Seriation Methods in Relation to a Family of Cluster Analysis Methods;
- Totally ordering the whole set of categories associated with a set of ordinal categorical attributes;
- Assignation problems in *Pattern Recognition* between geometrical figures where the quality measure of the assignation has to be independent of specific geometrical transformations applied on the figures concerned.

In this context, an additional chapter may be defined by the description and analysis of *Directed Binary Hierarchy*. The latter corresponds to an asymmetrical structure in which the leaf set is sustained by a linear order. To situate this structure—introduced by Régis Gras—with respect to the classical and then symmetrical binary hierarchy, it was formalized and studied in the first chapter of the Springer Nature book volume mentioned above, and previously, in an important article "Directed Binary Hierarchies and Directed Ultrametrics" by Israël César Lerman and Pascale Kuntz in *The Journal of Classification* (28): 272–296 (October 2011).

Resume in one book all of the themes that we have just quoted above would have required a too large volume whose material may appear as too scattered. In this book, we prefer to focus on the subject of *Seriation*. This corresponds to two of the chapters of the French book mentioned above. It is indeed a rather broad subject in its own right. On the other hand, the seriation approach has close links with Clustering (Classification). Both approaches have to be situated facing each other. This is the reason of the title of this book.

The works just mentioned have already given rise—apart from what is included in the french book—to a thesis [1] and the article [2]. The book proposed here goes very far beyond these works:

- The mathematical results are more accurate and richer;
- New and powerful algorithmic combinatorial and statistical methods are proposed and validated;
- A very rich synthesis is proposed, the new methods are situated in relation to a very important set of recent methods;
- A vast experimentation on simulated or real data is carried out.

Let us now briefly describe the respective contents of the different chapters.

The first chapter introduces the problem of seriation, the methods proposed to treat it, as well as the history of the evolution of the domain.

In the second chapter, a formal mathematical expression is proposed in the case where the data is defined by an incidence table composed of zeros and ones and therefore, where the descriptive attributes are Boolean. In this chapter, we establish the statistical bases leading to the development of an original and very simple methodology for planar geometric representation of the columns or rows of the incidence table. This method, called *Attraction Pole* method enables a seriation to be revealed. Its behavior will be experimented in Chap. 4.

Chapter 3 offers a detailed and commented description of the main methods considered in the literature to attack the seriation problem. The mutual links between the different approaches are brought out. This chapter ends by indicating a combinatorial version for attraction pole method. The latter will be developed in Chap. 5.

As just mentioned, Chap. 4 develops an experimental analysis comparing several methods of planar geometric representation: the new one of the poles of attraction and the classic and proven ones. For the latter, it is the principal component analysis, correspondence analysis and multidimensional scaling that underlies D. Kendall's Horse-shoe method (see [3]). Several seriation forms are tested, some of them include two or even three blocks.

In Chap. 5, a new family of ordinal and combinatorial seriation algorithms is established. In these, the seriation result depends on how to choose to start the order. The behavior of these algorithms is experimentally proven with respect to simulated or real data.

The determination of a system of attraction poles, mutually distant from each other enables a new and rich family of clustering methods to be derived. This is developed in Chap. 6. The respective relationships of this new family with that of the *K-means* on the one hand, and that of ascendant hierarchical classification on the other, are established.

In conclusion, we will give the possible extensions of the whole work.

Rennes, France   Israël César Lerman
Villetaneuse, France   Henri Leredde

References

1. H. Leredde, La méthode des pôles d'attraction, La méthode des pôles d'agrégation. PhD thesis, Université de Paris 6, October 1979
2. I. C. Lerman, H. Leredde, La méthode des pôles d'attraction, in *Analyse des Données et Informatique*, ed. by E. Diday et al. (IRIA, 1977), pp. 37–49
3. D. G. Kendall, Seriation from abundance matrices, in *Mathematics in Archaeological and Historical Sciences*, ed. by D. G. Kendall, F. R. Hodson, P. Tautu (Aldine-Atherton, Chicago, 1971), pp. 214–252

Acknowledgements

We acknowledge with gratitude the support of various institutions and individuals who have sponsored our work. The University of Rennes 1 and the University of Sorbonne Paris Nord have provided support in the framework of their respective departments the ISTIC (Informatique et Communications) and the Galilée Institute with Mathematics Department and his laboratory LAGA.

Gilles Lesventes (University of Rennes 1) and Jean-Pierre Astruc (President of the Sorbonne Paris Nord University) welcomed and supported us with great kindness.

Extensive facilities have been provided by the IRISA (Institut de Recherche en Informatique et Systèmes Aléatoire) Institute. We especially thank their directors Bruno Arnaldi, Jean Marc Jezequel and Guillaume Gravier.

Philippe Louarn (INRIA-Rennes) has defined the general LaTeX structure with respect to which we have composed this book. He helped us many times and his help was always valuable. We are very grateful to him.

We also have immense gratitude to the editor of the series in which this book is published as well to the editorial staff of Springer Nature, particularly to Helen Desmond who has always been in our listening.

Contents

Chapter 1
General Introduction: Methods and History

General Description The general framework of this book is unsupervised combinatorial and statistical data analysis. In this, the synthetic structure sought is discrete. Mainly, this structure can be in similarity classes or ordinal. Basically, whatever is the target structure, the information is the most often, provided by a data table crossing a set of objects with a set of descriptive attributes. This data structure refers to what is called "two-way two-mode data". Otherwise it happens that the data is directly provided by a symmetrical array of similarity (or dissimilarity) indices on the set **E** concerned by the analysis. This data structure refers to what is called "two-way one-mode data". The set **E** can be the object set or the attribute set.

Very generally, a method of data analysis uses the statistical approach to evaluate the links between pairs of elements of the same nature.

Compared to the formal structure mentioned "two-way two-mode data", these two elements can be attributes or objects. The attributes are represented in columns and the objects in rows of the data table.

Now, mathematically, in a fundamental way, a given method refers either to linear algebra by including diagonalization of a symmetric matrix, or to combinatorics, or even to a numerical optimization method which may have a combinatorial character.

If the mathematical tool is linear algebra or numerical optimization, the proposed synthesis structure is generally a planar geometric representation of the set concerned (set of attributes or set of objects in our case), by a cloud of points in such a way their mutual distances define an approximation of their mutual proximities.

If the mathematical tool is combinatorics, the synthesis structure is either symmetrical or asymmetrical. In the first case (symmetrical), the resemblance class structure is generally defined either by a partition or by an ordered partition chain (see Chap. 1 of [1]) on **E**. The latter case is characterized by cluster inclusions. More precisely, if C_1 and C_2 are two classes such that $C_1 \subset C_2$, it is desired that globally the mutual pairwise similarities are greater in C_1 than those in C_2.

In the asymmetrical case, the synthesis structure is defined by an order on **E**. This order is either total or partial. Most often, the ordinal structure on **E** is defined by a

I. C. Lerman and H. Leredde, *Seriation in Combinatorial and Statistical Data Analysis*, Advanced Information and Knowledge Processing, https://doi.org/10.1007/978-3-030-92694-6_1

total order. An ordinal structure is asymmetrical, whereas the classification structure is symmetrical. In a symmetrical structure if (x, y) is an ordered pair of elements, the role played by y with respect to x is the same as that of x with respect to y. This property does not hold for an ordinal structure as that considered in seriation problems. Most of this book is devoted to seriation but also in relation to classification (clustering) methods. New families of these will be proposed. As just indicated, the set $\mathbf{E}$ studied can be the object set or the attribute set.

The set $\mathbf{E}$ being provided with a similarity (resp., dissimilarity or distance) index, the objective consists of determining a total order on $\mathbf{E}$ such that, globally, the similarity between two elements of $\mathbf{E}$ is all the greater (resp., smaller) as these are ordinally close (resp., as that the difference between their ranks is large).

Mathematically, a seriation of $\mathbf{E}$ is a sequence of the $\mathbf{E}$ elements. This sequence defines a total order ω on $\mathbf{E}$ such that if (x, y) is an ordered pair of $\mathbf{E}$, $x\omega y$ if and only if x precedes y in the sequence. Clearly, a seriation defines a total order on $\mathbf{E}$ and vice versa a total order on $\mathbf{E}$ defines a seriation of $\mathbf{E}$. Both expressions will be used synonymously.

As expressed in [2] Seriation and Classification (Clustering) are intimately involved in a combinatorial and statistical data analysis. The methods referenced for this interaction are considered mainly in the case where the matter is to establish an order on the leaves of a hierarchical classification tree associated with similarities [3–7]. Note that classification and clustering will be used in our writing synonymously.

In order to make more explicit the interaction problems between clustering and ordering, let us consider a partition of $\mathbf{E}$ respecting mutual similarities. This partition can be noted in the form

$$\pi(\mathbf{E}) = \{E_1, E_2, \ldots, E_j, \ldots, E_k\} \tag{1.1}$$

where E_j denotes the jth class, $1 \leq j \leq k$.

Two seriation problems can be posed with respect to $\pi(E)$. These can be, respectively, qualified "within" and "between". For the first, it is the matter of building a seriation on each of the E_j sets. For the second problem, the ordering to be established is on the set of the classes, where each of them is considered as a single entity. In this case, we have to adopt a similarity or a dissimilarity measure on the class set. To this end, we can refer to Chap. 10 of [1] where several options are studied.

Some Historical Points Historically, according to Kendall [8, 9] the first introduction of formal methods in seriation is due to Flinders Petrie [10]. In accordance with the P. Ihm presentation [11], we will distinguish three periods in the evolution of formal methods for the treatment of seriation problems. Three main facets can be distinguished in these methods: algorithmic, geometrical, or combinatorial and statistical.

In the first period, it was necessary to convince the archeologist and the paleontologist of the actual contribution of formal methods in their fields. This period concluded with significant and decisive results of Ford and Wiley and Brainerd and Robinson and Ford and Elisséeff [12–15].

The second period was devoted to the development of methods resulting from the adaptation of general methods in multidimensional data analysis [16]. Two types of methods can be distinguished. The first one was worked extensively by Kendall [9]. It results from the adaptation of the $MDSCAL$ J.B. Kruskal method [17]. The second method is defined by the adaptation of correspondence analysis [11]. In a way, it can be said that the second period ended with two major conferences held in Marseille (France) and Mamaia (Romania):

- "Achéologie et Calculateurs", Eds J-C. Gardin and M. Borillo, Centre National de la Recherche Scientifique, 1970;
- "Mathematics in the Archeological and Historical Sciences" Eds F.R. Hodson, D.G. Kendall, FRS and P. Tàutu, The Royal Society and the Romanian Academy, 1970.

In our development, the problem of the formal search for a seriation is essentially considered as algorithmic, combinatorial and statistical. This seriation is defined by a complete order or total preorder on the row set or on the column set of a data table. The latter is either a presence/absence or percentage table. The use of a general method of data analysis rather than a direct method is a non-neutral choice. The advantage of a direct method is to be explicit in the progression of the algorithm concerned. Moreover, the first seriation methods associated with the authors mentioned above, proceeded directly by permuting rows and columns of the data table. This is the case, in particular, of Bertin works [4, 18–20] where visual perception is substituted for statistical approach. This is also the case of works such that those of Elisséeff or Robinson [21] and Brained [13, 15, 21–23]. Particularly, let us distinguish the Deutsch and Martin algorithm that we describe in Chap. 3, Sect. 3.2.5. In fact, the formalization of these latter contributions needs to be largely clarified, specified and systematized.

In the third period direct methods of grouping under constraints or combinatorial optimization appeared: Marcotorchino, Hubert, Caraux and Pinloche [24–26]. In these methods elaborate techniques are involved. By contrast, the original approaches presented in this book are based on very simple fundamental principles. Moreover, the algorithmic techniques associated—very simple and very direct—are justified from mathematical and statistical points of view.

New Contributions Let us take up again the mathematical expression of the seriation problem in the most often case of an incidence data table crossing a set $\mathcal{O}$ of objects in rows, with a set $\mathcal{A}$ of Boolean attributes in columns. The mathematical structure sought in seriation is an ordered pair of total preorders which correspond each other, respectively, on the sets of the row objects and column attributes. These two total preorders are obtained from a couple of permutations, subject to a statistical condition, on the row set and the column set. These permutations have to be determined in order to reveal a diagonal form called σ in the data table where the proportion of the $TRUE$ value of the Boolean attributes is significantly high. This objective is studied in Chap. 2. In this, seriation structure is formalized and original seriation methods are established.

In practice, most methods address the problem of the seriation of the set of rows representing objects. In our analysis, we consider two dual problems: that of seriation the attribute set and that of seriation the object set. In fact, we favor the option in which, first we build an ordering on the set of column attributes, and second derive an order on the set of row objects (see Chap. 5, Sect. 5.2).

Two types of original methods for establishing a seriation are considered in this book. The first one results from a planar Euclidean representation. This is done from a simultaneous analysis of the mean and variance of the similarities or (dissimilarities) of each of the elements to the other elements of the set concerned. In our case, this set can be the set $\mathcal{A}$ of the attributes or the set $\mathcal{O}$ of the objects (see Chap. 2, Sect. 2.4.5). Chapter 2 is devoted to the mathematical analysis of this approach. The results of this method, called geometrical "Attraction Pole" method, will be compared to the most classical factorial methods: principal component analysis and correspondence analysis. Moreover, we will consider multidimensional scaling approach results. This method sustains the Kendall Horse-shoe strategy (see Chap. 3, Sect. 3.5.2).

The second type of methods—developed in Chap. 5—is purely combinatorial. The seed idea of our combinatorial algorithms is simple. If we can determine the first element of the order defining the seriation, we can deduce the latter. In this type of approach, a set of ways to initialize a seriation is explored. This initialization can be performed as we have just said, with a single element, or with an ordered sequence of some elements. In these conditions, a given technique within this approach will be distinguished by the manner to obtain by successive chaining the ordered sequence of all the elements. For this purpose, an ordered sequence of elements is aggregated one by one, so that the last aggregated element is followed by its nearest successor or, more globally, by the nearest successor of all the aggregated sequence. In this approach, each element of the subset already seriated is compared to each element of the complementary subset not yet seriated. In these conditions, it is the distribution of pairwise similarities which makes it possible a rule in order to detect the next element of a given start of a seriation. Different versions of this rule are considered. This rule may use a geometrical planar representation that we just mentioned right above.

Faced with different solutions that can be obtained, it is a question of choosing the best or among the best ones. To this end, various criteria can be considered (see Chap. 3, Sects. 3.3.3 and 3.3.4). That expressed in Sect. 3.3.3 has a statistical significance.

Philosophically, as the case is for our approach, the Elisséeff heuristic (see Chap. 3, Sect. 3.2.4) refers to statistical and combinatorial considerations. However, the intuition which guides our strategy is very different. Moreover, the mathematical, statistical, algorithmic, and data processing development is much more systematic and coherent in our approach.

Moreover, let us point out that there is a conceptual difference in our method with respect to that of Elisséeff. In our technique, we initialize the process with respect to the origin of the order defining the seriation, whereas this initialization is made with respect to the median row of the data table in the Elisséeff heuristic.

Also, the algorithm of Deutsch and Martin [27] is combinatorial by nature, we describe it in Chap. 3, Sect. 3.2.5 and situate it in relation to our approach.

The intuitive logic of the formalization given by Robinson [21] also consists of playing a central role at the level of the median row of the data table. However, in this formalization, the data table provided by anthropology is not a presence/absence table. In this different data table, each row is defined by an empirical probability distribution on a set of types of an archeological object, such, for example, as a pottery.

A row in the latter data table corresponds to the description of a deposit. It corresponds to a stratum of a given trench. If st1 and st2 are two strata of the same trench such that st1 is above st2, then the deposit found in st1 is chronologically more recent than that found in st2. In these conditions, it is a question of mutually ordering the deposits found in the different trenches, that is, to determine a global order constrained by the local orders associated with the strata of each trench.

Two deposits located in two distinct trenches are considered all the more contemporary as the similarity between their respective profiles is great. It is from this consideration that an algorithmic search method of a global order can be built. The technique worked on relies on permutation of rows and columns of the symmetric similarity matrix between the deposits [21]. Previously, it is checked that this arrangement puts in the right order the deposits of the same trench in accordance with the layout of the strata.

Each deposit is defined as an aggregate of objects. The latter aggregate is described by a categorical attribute for which the attribute states are the different types. Thereby, each deposit is represented by the relative frequency distribution of the different attribute states on the deposit. More precisely, the value of the relative frequency associated with a given attribute state, which we denote by t, is defined as the proportion of objects in the deposit for which the value of the categorical attribute is t. Therefore, the similarity coefficient between two deposits has to be conceived for comparing empirical probability distributions on a finite set (the set of types) (see in [1], Chap. 7, Sect. 7.2.3).

We will not deal in this book with the data structure, just described and which comes from anthropology. We will mainly deal with Boolean data associated with a presence/absence data table crossing a set of objects with a set of Boolean attributes. Nevertheless, the results obtained in this case will contribute to the analysis of anthropological data.

The fundamental idea of Robinson [21] is relative to the characterization of the ordinal structure associated with a seriation of a set $\mathbf{E}$ endowed with a similarity (resp., dissimilarity) index, at the level of the symmetrical similarity or dissimilarity matrix on $\mathbf{E}$. When seriation underlies the data, a joint permutation of rows and columns reduces the matrix to a characteristic form. However, in the development given in [21], the expression of the similarity between deposits is not clearly defined. As just expressed, for this similarity, empirical probability distributions on the types associated with the deposits must be compared.

The Robinson condition (see 3.79 and 3.80 in Sect. 3.3.2 of Chap. 3) is satisfied for the mathematical formulation of a seriation in the case of an incidence (presence/absence) data table crossing objects in rows with Boolean attributes in columns (see Sect. 2.2 of Chap. 2). The Robinson form is relative to the symmetrical matrix of association coefficients (see 2, Sect. 2.3.3) between the elements of the set concerned arranged according to the seriation order.

A seriation structure may correspond to more than a single homogeneous block (see Sects. 2.2 and 2.4.3 of Chap. 2). This configuration has already been considered in [28]. In this case, the different blocks follow each other in the order corresponding to the global seriation. The elements of each of the blocks (column attributes in our formalization) are ordered according to the global seriation. This situation has already been presented at the beginning of this introduction, where the role of bringing out the blocks is considered as reserved for a clustering technique.

The "Block seriation" method of Marcotorchino [24] is, in fact, a constrained bi-classification (bi-clustering) method of the row and column sets of an incidence data table (see Chap. 3, Sect. 3.2.2). Let us stress here that our processing data—underlying seriation of blocks—is based on the new methods proposed in this book. These can refer to the geometric representation of the Attraction Pole method (see Chap. 2, Sect. 2.4.5) or more effectively to combinatorial and statistical algorithms (see Chap. 5).

As already emphasized at the beginning of this introduction, ordinal and clustering methods are intimately involved in a combinatorial and statistical data analysis. The two main types of approaches (clustering and ordering) can be methodologically associated or derived from independent techniques.

In Chap. 2, we show how the determination of the attraction poles—provided by a simultaneous analysis of the mean and variance of proximities from each element to the other ones—allows a geometrical representation plane to be established. Such an analysis can be directed towards the development of a new family of clustering algorithms which operates in successive divisions around attraction poles ([29, 30] see Chap. 6). This method contributes in an interesting way to the fundamental problem of initialization of the K-means algorithm [31, 32].

In these methods, two interpretations of the notion of proximity or difference can be considered. If it is a matter of organizing a set of descriptive attributes, the notion of statistical independence can prevail. The notion of distance or dissimilarity can also be considered for such an organization. However, the latter notion intervenes more appropriately for the organization by the classification (clustering) of the object set.

Comparing Methods One of the major interests of the Liiv article [2] consists of reviewing a large scope of application domains of seriation (Archeology and Anthropology, Cartography and Graphics, Sociology and Sociometry, Psychology and Psychometrics, Ecology, Biology and Bioinformatics, Group Technology and Cellular Manufacturing, Operations Research). In these, graphical representation of the seriation takes an important part. The formal introduction and development of seriation techniques were motivated by these fields of application research and more specifically by archeology and anthropology.

The description of these techniques as given in [2] is too succinct and subjective, so that it is difficult to understand their respective mechanisms. On the other hand and in a related way, the general graph given of derivation of the different methods, one compared to the others, is questionable. We prefer to give in Chap. 3 an explicit enough description of the main mutually different methods proposed. This enables the reader to understand deeply the different approaches and their mutual relationships. This description is organized as follows:

- Visual and combinatorial methods;
- Combinatorial optimization methods;
- Spectral methods;
- Methods using planar geometrical representation;
- Algorithmic and combinatorial methods.

To finish, even at the risk of repeating ourselves, let us situate the material of this book in relation to those published and more precisely to [33, 34].

The book [33] is devoted to time gripping in the analysis of archeological phenomena. In this book, Fig. 3.1 provides an interesting taxonomy of seriation techniques. In this, our development refers to similarity seriation or *seriation by resemblance*. Compared to the book [34], our development is much more precise and complete on the following points:

1. Mathematics;
2. Statistics;
3. Overall view;
4. Experimental validation.

In this way, new methods can be discovered and analytical comparison methods can be performed.

As indicated above, we focus on the data case structure provided by a rectangular data table composed of zeros and ones, crossing a set of objects in rows with a set of Boolean attributes in columns (two-way two-mode data). This case is most frequent in the literature concerned. The formalization of this data table allows a precise and new mathematical definition of seriation.

The establishment of a statistical association coefficient between Boolean attributes observed on a set of objects gives rise to two symmetrical similarity matrices (two-way one-mode data). The first matrix is between the Boolean attributes, and the second between the described objects. Analytically, a distance index can naturally be associated with the similarity index. In these conditions, we can work either with a matrix of similarities, or with a matrix of distances.

This construction makes it possible development of new methods in geometric or combinatorial representations of seriation and also importantly in classification (clustering). These methods are based mathematically, statistically and are discussed in relation to other methods. This in relation to processing simulated or real data. In the literature, the problems of data formal representation and those of building similarities or distances are not really taken into account. In addition, the logical or statistical natures of the criteria defined to be optimized for the seriation search are established a priori.

In combinatorial optimization, these criteria may relate to a matrix of dissimilarity (resp., distance) between the elements of the set concerned, or more locally, to the Objects x Attributes incidence data matrix. More specifically, as already expressed an entire chapter of our book (Chap. 3) gives an original overview of a wide range of seriation methods including different types of criteria. In the case where a given method uses a matrix of similarities or dissimilarities, resulting from a table of data description, our approach allows generalization to any type of data table description (see [1]).

However, in this generalization, only categorical attributes can be substituted for Boolean ones. In these conditions, the value of a given attribute on a given object is unique in relation to the scale of the categorical attribute concerned (see Chap. 3 of [1]).

The generalization to a description as given in [35, 36] by attributes defining complex structures is difficult to manage, for interpretative analysis.

It is much more difficult and especially uncertain in the case of the search for a seriation in the case of a description called "symbolic" (see [37]). The starting point is classic. It is situated in relation to a potential representation space where a given variable has a given value on an indivisible object. However, from such a description by initial variables, Boolean description attributes are introduced, which are, respectively, called *symbolic* objects. Each of such attributes is defined from a logical formula which can be complex, relating to all the values of the initial variables.

One last technical point. The reader is assumed to be familiar with the intuitive notion of similarity index and with various formal expressions related to the mathematical representation of the items to be compared. We will use as synonym expressions like index and coefficient on the one hand, similarity, proximity and association, on the other hand. The notion of index of dissimilarity is axiomatically more general than that of index of distance. In addition, as above, we will use the terms classification and clustering as synonyms.

References

1. I.C. Lerman, *Foundations and Methods in Combinatorial and Statistical Data Analysis and Clustering* (Springer Nature, 2016)
2. I. Liiv, Seriation and matrix reordering methods: An overview. Wiley InterScience, pp. 70–91 (2010), www.interscience.wiley.com, 10.1002
3. G. Gruvaeus, H. Wainer, Two additions to hierarchical cluster analysis. Br. J. Math. Stat. Psychol. **25**, 200–206
4. J. Bertin, Traitements graphiques et mathématiques. différence fondamentale et complémentarité. Mathématiques Sci. Hum. **80**, 60–71 (1980)
5. G. Brossier, Problèmes de représentation par des arbres, Thèse de Doctorat ès Sciences. Ph.D. thesis, Université de Rennes 2, June 1986

6. Z. Bar-Joseph, D.K. Gifford, T.S. Jakkola, Fast optimal leaf ordering for hierarchical clustering. Bioinformatics **17**(S1), S22–S29 (2001)
7. Z. Bar-Joseph, E.D. Demaine, D.K. Gifford, N. Srebro, A.M. Hamel, T.S. Jakkola, K-ary clustering with optimal leaf ordering for gene expression data. Bioinformatics **19**(9), 1070–1078 (2003)
8. D.G. Kendall, A statistical approach to flinders petrie's sequence-dating. Bull. Int. Stat. Inst. **40**, 657–681 (1964)
9. D.G. Kendall, Seriation from abundance matrices, in *Mathematics in Archaeological and Historical Sciences*, ed. by D.G. Kendall, F.R. Hodson, P. Tautu (Aldine-Atherton, Chicago, 1971), pp. 214–252
10. W.M.F. Petrie, Sequences in prehistoric remains. J. Anthropol. Inst. **29**, 295–301 (1899)
11. P. Ihm, A contribution to the history of seriation in archaeology, in *Studies in Classification, Data Analysis and Knowledge Organization, Proceedings of the 28th Annual Conference of the Gesellschaft für Klassifikation*, ed. by W. Gaul, C. Weihs, March 9-11, 2004, pp. 307–316. University of Dortmund (2005)
12. J.A. Ford, G.R. Willey, Surface survey of the virù valley. Anthropol. Pap. Am. Mus. Nat. Hist. **43**, 1–89 (1949)
13. G.W. Brainerd, The place of chronological ordering in archaeological analysis. Am. Antiq. **16**, 301–313 (1951)
14. J.A. Ford, A quantitative method for deriving cultural chronology. Pan American Union, Washington, D.C. (1962)
15. V. Elisséeff, Possibilités du scalogramme dans l'étude des bronzes chinois. Mathématiques Sci. Hum. **11**, 1–10 (1964)
16. H. Leredde, F. Djindjian, Traitement automatique des données en archéologie. Dossiers de l'Archéologie **42**, 52–69 (March 1980)
17. J.B. Kruskal, Multi-dimensional scaling. Psychometrika **29**, 1–27, 28–12 (1964)
18. J. Bertin, *Sémiologie graphique : les diagrammes, les réseaux, les cartes* (Mouton, Paris)
19. J. Bertin, *La Graphique et le traitement graphique de l'Information* (Flammarion, 1977)
20. J. Bertin, *Graphics and Graphic Information Processing* Translated by William J. Berg and Paul Scott (Walter de Gruyter, Berlin)
21. W.S. Robinson, A method for chronologically ordering archaeological deposits. Am. Antiq. **16**, 293–301 (1951)
22. V. Elisséeff, De l'application des propriétésdu scalogramme à l'étude des objets. Calcul et formalisation dans les sciences de l'homme, pp. 107–120 (1968)
23. V. Elisséeff, Données de classement fournies par les scalogrammes privilégiés, in *Archéologie et Calculateurs*, ed. by M. Borillo, J.-C. Gardin, pp. 177–180. Centre National de la Recherche Scientifique (1970)
24. F. Marcotorchino, Block seriation problems: a unified approach. Appl. Stoch. Model. Data Anal. **3**, 73–91 (1987)
25. L.J. Hubert, Some applications of graph theory and related nonmetric techniques to problems of approximate seriation: The case of symmetric proximity measures. J. Math. Stat. Psychol. **27**, 133–153 (1974)
26. G. Caraux, S. Pinloche, Permutmatrix: a graphical environment to arrange gene expression profiles in optimal linear order. Bioinformatics **21**(7), 1280–1281 (2005)
27. S.B. Deutsch, J.J. Martin, An ordering algorithm for analysis of data arrays. Oper. Res. **19**(6), 1350–1362 (1971)
28. I.C. Lerman, Analyse du phénomène de la "sériation" à partir d'un tableau d'incidence. Mathématiques Sci. Hum. **38**, 39–57 (1972)
29. H. Leredde, La méthode des pôles d'attraction, La méthode des pôles d'agrégation. Ph.D. thesis, Université de Paris 6, October 1979
30. I.C. Lerman, *Classification et analyse ordinale des données* (Dunod, 1981)
31. I.C. Lerman, Convergence optimale de l'algorithme de "réallocation-recentrage" dans le cas continu le plus simple. R.A.I.R.O. Recherche opérationnelle/Oper. Res. **20**(1), 19–50 (1986)

32. G. Celeux, Étude exhaustive de l'algorithme de réallocation-recentrage dans un cas simple. R.A.I.R.O. Recherche opérationnelle/Oper. Res. **20**(3), 229–243 (1986)
33. M.J. O'brien, L.R. Lyman, *Seriation, Stratigraphy, and Index Fossils: The Backbone of Archaeological Dating* (Kluwer Academic/Plenum, 1999)
34. D.L. Carlson, *Quantitative Methods in Archeology Using R* (Cambridge Unuversity, 2017)
35. I.C. Lerman, Ph. Peter, Representation of concept description by multivalued taxonomic preordonance variables, in *Selected Contributions in Data Analysis and Classification*, ed. by P. Brito, F. Carvalho, G. Cucumel (Springer, 2007), pp. 271–284
36. I.C. Lerman, Ph. Peter, Indice probabiliste de vraisemblance du lien entre objets quelconques : analyse comparative entre deux approches. Rev. Stat. Appliquée **LI**(1), 5–13 (2003)
37. H.H. Bock, E. Diday (ed.), *Analysis of Symbolic Data* (Springer, 2000)

Chapter 2
Seriation from Proximity Variance Analysis

2.1 Introduction

In order to resolve chronological problems, *Seriation* was introduced by archeologists as a formal algorithmic method. According to [1], Petrie (1899) [2] was the first to propose such an approach.

Liiv [3] provides with 171 references an almost exhaustive review of how seriation was defined, invented and mainly applied in fields as different as archeology and anthropology, cartography and graphics, sociology and sociometry, psychology and psychometrics, ecology, biology and bioinformatics, group technology and cellular manufacturing and operations research. However, to cover all these contexts the definition proposed in [3] is very general without a precise mathematical statement.

For our development, to begin, let us formulate in an intuitive manner the problem concerned by considering an example in archeology. The data is defined by a set $\mathcal{O}$ of n tombs or graves described by a set $\mathcal{A}$ of p rites. Each of the latter is supposed to be formalized by a Boolean attribute. Therefore, an incidence data table represents the data description (see Chap. 3 of [4]). By denoting $\mathcal{O}$ and $\mathcal{A}$, respectively, as

$$\mathcal{O} = \{o_i | 1 \leq i \leq n\}$$

and

$$\mathcal{A} = \{a^j | 1 \leq j \leq p\} \tag{2.1}$$

The data table $\mathcal{T}$ can be written as follows:

$$\mathcal{T} = \{x_i^j | 1 \leq i \leq n, 1 \leq j \leq p\} \tag{2.2}$$

where $(\forall (i, j), 1 \leq i \leq n, 1 \leq j \leq p)$, $x_i^j = 1$(resp., 0) if the ritual a^j is present (resp., absent) in the tomb o_i.

As time passes various rites succeed. A given grave does correspond to an age even more remote that it contains the most ancient rites. In these conditions, the

© The Author(s), under exclusive license to Springer Nature Switzerland AG 2022
I. C. Lerman and H. Leredde, *Seriation in Combinatorial and Statistical Data Analysis*, Advanced Information and Knowledge Processing, https://doi.org/10.1007/978-3-030-92694-6_2

objective consists of discovering the chronological order on $\mathcal{A}$ and the associated one on $\mathcal{O}$.

The archeologist hypothesis $\mathcal{H}$ which makes possible a solution for this problem can be stated as follows:

A given ritual is practiced for a continuous period of time; moreover, if a^{t_1}, a^{t_2} *and* a^{t_3} *are three rites appeared on the dates* t_1, t_2 *and* t_3, *where* $t_1 < t_2 < t_3$, a^{t_1} *is closer to* a^{t_2} *than to* a^{t_3}, *on the other hand,* a^{t_2} *is closer to* a^{t_3} *than* a^{t_1} *to* a^{t_3}.

A mathematical model of seriation will be defined in the next section (see Sect. 2.2). This has to be considered as approximating the reality. Some rituals may disappear and reappear through the ages. Also in paleontology (see [5]), there is noise in real data: false positive (false values 1) and false negative (false values 0) may occur. Nevertheless, in real data, the truthfulness of the hypothesis $\mathcal{H}$ is supposed to be strong enough in order to be revealed statistically.

As a matter of fact, discovering seriations is a general problem in *Behavioral Sciences* (see, for example, [3]) where a fundamental purpose consists of detecting from the data phenomena which evolve and discovering variables which control such evolutions.

The focus of attention of most seriation methods is ordering first the object set $\mathcal{O}$. In our approach, given an incidence data table, the first objective consists of ordering the attribute set $\mathcal{A}$ represented by the columns of the data table. In these conditions, the row data table ranking associated with the seriation model is derived in a second step. In Sect. 5.2 of Chap. 5, a process will be specified in order to infer the latter row ranking from the column ranking. Nevertheless, the new methods developed in this book make possible to directly determine a seriation on the object set.

The method we consider for ordering the attribute set $\mathcal{A}$ is based on the distribution of the association coefficients between the Boolean attributes indexing the columns of the incidence data table. And, in this respect, the association coefficient we will adopt, established in Sect. 5.2 of Chap. 5 of [4], neutralizes in the association between two attributes, the variation influence of the number of objects where a given attribute is present. This coefficient will be recalled in Sect. 2.3.2.

In the latter section, it is shown that in the case for which the mathematical model of seriation holds, the respective columns of the incidence data table are representable isometrically by points of a 1 length oriented line segment of the axis associated with the reals.

In Sect. 2.4.2, an optimal property is established relative to this representation. This point may be the extreme left or the extreme right one. The latter property is expressed with respect to the proximity distribution of the point concerned with the other points of the representation.

A general formula of proximity variance analysis is proposed in Sect. 2.4.4. In this section, we show how simultaneous mean and variance analysis of mutual proximities enable more than one seriation block to be revealed. This leads to a synthetic plane representation for the columns of the incidence data table (see Sect. 2.4.5). In fact, this representation has the same conceptual nature as that provided by principal component factorial analysis.

Chapter 3 is devoted to outline description of classical seriation methods. With respect to these, the specificity of our approach will be highlighted.

In this chapter, the contributions provided by [6–8] and Chap. 8 of [9] are worked on again with a certain hindsight. Besides, let us indicate that the first theoretical contribution of the substantive and important thesis [8] is [6].

2.2 Definition of a Seriation; Its Unicity

2.2.1 Definition of a σ Form

Suppose the above archeologist hypothesis $\mathcal{H}$ exactly verified. By assuming the tombs chronologically ordered in the incidence data table, for any index j, $1 \leq j \leq p$, there exists two row indices $i_1(j)$ and $i_2(j)$, $1 \leq i_1(j) \leq i_2(j) \leq n$, such that

$$(\forall i), i < i_1(j) \text{ or } i > i_2(j), x_i^j = 0 \text{ or not defined}$$
$$\text{and}$$
$$(\forall i), i_1(j) \leq i \leq i_2(j), x_i^j = 1 \tag{2.3}$$

Now, suppose the rites—which label the columns of the incidence data table—ranked according to the respective dates where they appeared, if the expert assumption is exactly verified, for each row index i, $1 \leq i \leq n$, there exists two column indexes $j_1(i)$ and $j_2(i)$, $1 \leq j_1(i) < j_2(i) \leq p$, such that

$$(\forall j), j < j_1(i) \text{ or } j > j_2(j), x_i^j = 0 \text{ or not defined}$$
$$\text{and}$$
$$(\forall j), j_1(i) \leq j \leq j_2(i), x_i^j = 1 \tag{2.4}$$

Notice that only (2.3) is considered in [5] where seriation problem is addressed in paleontology. In the case where the latter equation holds, according to the terminology adopted in [5], *Lazarus events* do not exist. In fact this condition corresponds exactly to the "Consecutive 1 Property" ($C1P$) expressed in [10]. In our case, we will be first concerned by the (2.4) formulation. This is because, as mentioned in the introduction, the first step of our approach consists of finding an ordering on the rite set (the attribute set). For this order, each satisfies (2.4) and then an order on the row table can be derived (see Chap. 5, Sect. 5.2).

To define the seriation form σ, the following condition C_0 is supposed to be satisfied.

C_0: There does not exist two distinct attributes (rites) which are simultaneously present or absent of each of the objects (tombs) described.

In other words, the respective columns of the incidence data table are mutually different. This condition does not entail any generality restriction of the problem of establishing a seriation structure. Indeed, it is easy to come down to the case for which condition C_0 holds by describing the incidence data table column by column and by deleting each column vector already met. On the other hand, it will be shown by the mathematical analysis that column repetition does not provide any additional data for the solution of the seriation problem.

The following condition C_1 seems to be very restrictive.

C_1: The number of objects possessing a given attribute is the same for each of the descriptive attributes.

In other terms, the number of 1 component per column vector is the same for the different columns. Considered in archeology, this property means that the number of tombs where a given rite appears is constant whatever the latter. In other words, the life lengths of the different rituals are equal. This condition may not be considered as acceptable by archeologists. However, it makes possible mathematical analysis of the seriation structure. On the other hand and above all, the approach emphasized is based on the distribution of the association coefficients between the Boolean attributes which label the columns of the incidence data table. And, in this respect, the association coefficient we adopt (see Sect. 5.2 of Chap. 5 of [4]) neutralizes the variation influence of the number of objects for which a given attribute is present. This coefficient will be recalled in Sect. 2.3.2.

Figure 2.1 illustrates the σ form of an incidence data table, for which the conditions C_0 and C_1 are satisfied.

Once again, let us emphasize that the absence of Lazarus events [5] corresponds to the satisfaction of the $C1P$ (consecutive ones property) as defined in [10]. In our

Fig. 2.1 σ form of the incidence data table

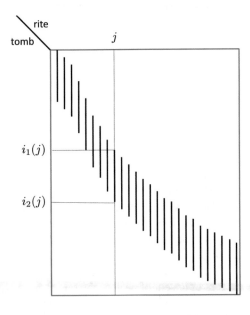

definition of the *sigma* form, the latter condition holds. In addition, for the *sigma* form the number of ones per column is constant.

2.2.2 Unicity

With the condition C_1, Eq. (2.3) becomes

$$(\forall i), i < i_1(j) \text{ or } i > i_1(j) + t - 1, x_i^j = 0 \text{ or not defined}$$
and
$$(\forall i), i_1(j) \leq i \leq i_1(j) + t - 1, x_i^j = 1 \tag{2.5}$$

where for all j, $1 \leq j \leq p$, $t = i_2(j) - i_1(j) + 1$ is the constant number of objects (tombs) for which $x_i^j = 1$.

Now, consider the following mapping denoted by c:

$$(\forall j), 1 \leq j \leq p, j \mapsto c(j) = i_1(j)$$

c increases strictly and we have $c(1) = 1$ and $c(p) + t - 1 = n$.

A homothetic mapping γ can be associated with c as follows:

$$(\forall j), 1 \leq j \leq p, \ \gamma(j) = \frac{c(j)}{n}$$

We have:

$$\gamma(1) = \frac{1}{n} \text{ and } \gamma(p) = 1 - \frac{t-1}{n}$$

Definition 1 *The σ form is called reduced to a σ parallelogram form if and only if for each index column ordered pair (k, l) $(k < l)$, we have*

$$\frac{\gamma l - \gamma k}{l - k} = \alpha$$

where α is a constant.

Definition 2 *The incidence data table is called strongly vertically chained if and only if for each index column ordered pair (k, l) $(k < l)$, we have*

$$s(k, l) = \sum_{1 \leq i \leq n} x_i^k x_i^l > 0$$

This condition means that for each attribute pair, there exists at least one object for which both attributes k and l are *TRUE*.

Definition 3 *The incidence data table is called weakly vertically chained if and only if for each index column ordered pair (k, l) $(k < l)$, there exists a column sequence $(k_1, k_2, ..., k_r)$, such that $k_1 = k$, $k_r = l$ and*

$$min\big(s(k_1 k_2), s(k_2 k_3), ..., s(k_{r-1} k_r)\big) > 0.$$

Clearly, an incidence data table strongly vertically chained is, *a fortiori*, weakly vertically chained. In the case for which the latter condition is not satisfied, the data table is said to be *disconnected*. In this case, a *connected* component of the column vector set is defined as a maximal subset of column vectors such that the restriction of the data table to the latter subset is weakly vertically chained.

Notice that in the case of Fig. 2.1, the incidence data table is *weakly* vertically chained.

Proposition 1 *An incidence data table can be reduced to a σ form if and only if there exists a permutation of the data table rows such that entries comprising the value 1 of each column determine a row interval.*

This permutation condition is supposed to be satisfied in the definition given in (2.3). Clearly, the condition considered is necessary. It is also sufficient. The column order corresponds to the increasing sequence of the $\gamma(j)$ values, $1 \le j \le p$.

Note that to a σ form corresponding to (2.3), given by a row permutation τ, an analogous σ form can be associated bijectively with the reverse of τ. In practice, it is the knowledge of the expert which enables the right version between τ or its reverse to be considered.

For a row permutation τ determining a σ form of the incidence data table, the column order (ranking) is established in a unique way. Each column defines on the row set a total preorder into two or three classes (two classes for the first and last columns and three classes for the others). In the case of three classes, the median one corresponds to a vertical segment of the column concerned whose entries include the 1 value. The two other classes correspond to vertical segments, whose entries include the 0 value. In the case of the first (resp., last) column, two classes are defined. The first (resp., the second) corresponds to a vertical segment whose entries include the 1 value. The second class (resp., the first class) corresponds to a column vertical segment whose entries include the 0 value (see Fig. 2.1). The intersection of the different total preorders on the row set determines on the latter set a total preorder— that we denote by ν—with p classes (p is the column number). By saying that a given row is $TRUE$ (resp., $FALSE$) for a given column if the value 1 (resp., 0) is included at their intersection entry, a ν preorder class is composed of all rows such as for every column, they are simultaneously either $TRUE$ or, exclusively, $FALSE$.

A row permutation τ defines a total order—that we denote by ω—on the row set. τ is said *compatible* with the total preorder ν if

$$[\forall (h, i), 1 \le h, i \le n, h < i \text{ for the quotient order defined by } \nu] \Rightarrow [h < i \text{ for } \omega]$$

Now, let ν' be a ν reverse total preorder on the row set. Clearly, every row permutation compatible with ν or, exclusively, with ν' gives a σ form to the incidence data table. Additionally, each of two permutations compatible both with ν (or with ν') can be deduced each from the other by a product of local permutations operating each on a ν preorder class. Obviously, we can state

Proposition 2 *If an incidence data table can be reduced to a σ form vertically weakly chained, the only row permutations which preserve the σ form are compatible with ν or with ν'.*

For a σ form, the ν preorder classes and the associated quotient order, up to its reverse, are determined in a unique fashion. Effectively, an extreme ν preorder class can be defined as a maximal row set whose elements are simultaneously true in exactly one column. This extreme class defines the first column of the data table. From the latter, the other columns are established one by one. The jth column is that, among the non-selected columns, for which the number of rows simultaneously true in the $(j-1)$th and the jth columns is maximum. The column ranking obtained enables the total preorder ν or ν' to be determined.

To conclude, if an incidence data table admits a σ form reduction, the column total order which permits us to set up it is built up to a reverse order one. On the other hand, following Proposition 2, from a given σ form, all the row permutations preserving it can be enumerated.

2.3 Association Coefficient Between Columns of a σ Seriation Form

2.3.1 Preamble

We have just seen above that when an incidence data table can be reduced to a σ form satisfying conditions C_0 and C_1 (see (2.3) and above), the column order enables the associated row order to be retrieved.

As expressed in the Introduction (see Sect. 2.1) the σ mathematical definition corresponds to an idealization of seriation phenomena as they appear in real cases. For the latter, we have in fact to determine the column order which gives rise to a statistical profile of the σ form. This might be intuitively defined by (2.3) but where equations such as "$x_i^j = 0$" and "$x_i^j = 1$" have to be replaced by expressions such as "$x_i^j = 0$, the more often than not" and "$x_i^j = 1$, the more often than not", respectively.

For this statistical objective, our method will be based on the mutual analysis of the relationships between the Boolean attributes labeling the respective columns of the incidence data table. Therefore, the association coefficient which will be used in order to compare the column attributes has to be specified (see Sect. 2.3.2). As indicated in the Introduction, this coefficient is that established in Sect. 5.2 of Chap. 5 of [4]. It has the property to neutralize in a certain extent the rigor of the C_1 condition.

This coefficient will be taken up again in Sect. 2.3.2. It will be detailed in the case of a σ form defined mathematically (see Sect. 2.3.2). Besides, all the analysis which follows in Sects. 2.3.3, 2.3.4 and 2.4 is considered in the case of a σ form defined exactly. However, the methods proposed in the latter sections apply to real cases in order to discover a statistical σ form underlying the incidence data table.

2.3.2   The Association Coefficient

Let us return for a moment to the incidence data table

$$T = \{x_i^j | 1 \leq i \leq n, 1 \leq j \leq p\}$$

(see (2.2)) in order to recall its different formulations. This data table is used here to cross a set $\mathcal{A}$ of Boolean attributes with a set $\mathcal{O}$ of objects. The columns (resp., the rows) of this table represent the attributes (resp., the objects). As exemplified above in the archeology context, $\mathcal{A}$ is a set of rites and $\mathcal{O}$ a set of tombs. $x_i^j = 1$ (resp., $x_i^j = 0$) if the jth attribute which we denote by a^j is present (resp., absent) in the ith object denoted by o_i, $1 \leq i \leq n, 1 \leq j \leq p$. With each element a^j of $\mathcal{A}$ is associated a logical vector column that we denote by x^j, expressed as follows:

$$x^j =^T (x_1^j, x_2^j, ..., x_i^j, ..., x_n^j) \tag{2.6}$$

where the exponent T indicates *Transpose*.

Thus, x^j is a summit of the logical cube $\{0, 1\}^n$, $1 \leq j \leq p$. The latter defines one version of the attribute representation space.

Otherwise, x^j can be assimilated to an indicator vector of an $\mathcal{O}$ subset $\mathcal{O}_j$ including all of the objects for which the attribute a^j is $TRUE$. In these conditions, an equivalent version of the representation space is defined by the set $\mathcal{P}(\mathcal{O})$ of all $\mathcal{O}$ subsets.

Thereby, the incidence data table T provides us the attribute set $\mathcal{A}$ as a sample of p points (resp., elements) in $\{0, 1\}^n$ (resp., in the set $\mathcal{P}(\mathcal{O})$ of all $\mathcal{O}$ parts).

We may also regard the incidence data table T more symmetrically and graphically as a distribution of 0 and 1 weights on the rectangle $[1, 2, ..., i, ..., n] \times [1, 2, ..., j, ..., p]$ of $\mathbb{N}^2$, where $\mathbb{N}$ is the set of integers.

The construction of an association coefficient between two column attributes a^k and a^l, $1 \leq k < l \leq p$, refers to the former representation in $\{0, 1\}^n$ (resp., in $\mathcal{P}(\mathcal{O})$). The more graphical and less spatial representation will better guide our intuition for seeking a seriation.

As just mentioned above, the association coefficient we consider between the Boolean attributes a^k and a^l, $1 \leq k < l \leq p$, was established in Sect. 5.2 of Chap. 5 of [4]. The starting point of this construction is the *raw* similarity coefficient

$$s(a^k, a^l) = card(\mathcal{O}_k \cap \mathcal{O}_l) = \sum_{1 \leq i \leq n} x_i^k . x_i^l \tag{2.7}$$

where—let us repeat it—$\mathcal{O}_j$ designates the subset of $\mathcal{O}$ constituted of all objects on which the Boolean attribute a^j is $TRUE$.

$s(a^k, a^l)$ is normalized statistically with respect to a probabilistic hypothesis of no relation $\mathcal{H}$. For this purpose, an ordered of *independent* pair (X_k, X_l) is associated with $(\mathcal{O}_k, \mathcal{O}_l)$. X_k and X_l are random subsets of a finite set, such that the mathematical expectations $\mathcal{E}[card(X_k)]$ and $\mathcal{E}[card(X_l)]$ are equal to $card(\mathcal{O}_k)$ and $card(\mathcal{O}_l)$, respectively.

Three fundamental forms $\mathcal{H}_1$, $\mathcal{H}_2$ and $\mathcal{H}_3$ of the hypothesis of no relation $\mathcal{H}$ emerge. They correspond for the probability law of $s^\star(k, l) = card(X_k \cap X_l)$ to hypergeometric, binomial and Poisson models, respectively (see the reference above in [4]).

The general form of the normalized coefficient is

$$S(a^k, a^l) = \frac{s(a^k, a^l) - \mathcal{E}[s^\star(k, l)]}{\sqrt{var[s^\star(k, l)]}} \tag{2.8}$$

It is the $\mathcal{H}_3$ version of the hypothesis of no relation which is retained here. In fact, it turns out to be the more relevant one for comparing Boolean attributes whose respective presences in the objects described are not very frequent. Moreover, $\mathcal{H}_3$ leads to the simplest mathematical expression for the normalized association coefficient $S(a^k, a^l)$ (see (2.8)). The coefficient obtained designated by Q_3 in (5.2.32) of Sect. 5.2 of Chap. 5 of [4] becomes here

$$S(k, l) = \frac{s(k, l) - n.\eta_k.\eta_l}{\sqrt{n.\eta_k.\eta_l}} \tag{2.9}$$

where $s(k, l)$ and $S(k, l)$ stand for $s(a^k, a^l)$ and $S(a^k, a^l)$, respectively. On the other hand, η_k (resp., η_l) designates the proportion (relative frequency) of objects in $\mathcal{O}$ for which a^k (resp., a^l) is $TRUE$, namely,

$$\eta_k = \frac{card(\mathcal{O}_k)}{n} \text{ and } \eta_l = \frac{card(\mathcal{O}_l)}{n} \tag{2.10}$$

Remember that under the $\mathcal{H}_3$ hypothesis, the random index

$$S^\star(k, l) = \frac{s^\star(k, l) - n.\eta_k.\eta_l}{\sqrt{n.\eta_k.\eta_l}} \tag{2.11}$$

follows in the usual conditions ($n.\eta_k.\eta_l$ not too small) a very near standard normal distribution $\mathcal{N}(0, 1)$.

Now, consider the following substitution

$$\left(\forall (i, j), 1 \leq i \leq n, 1 \leq j \leq p, \right), x_i'^j = \frac{x_i{}^j - \eta_j}{\sqrt{\eta_j . \sqrt{n}}} \longrightarrow x_i'^j \qquad (2.12)$$

in the incidence data table entries. By applying it to $s(k, l)$ (see (2.7)), we obtain $S(k, l)$ (see (2.9)).

$$S(k, l) = \sum_{1 \leq i \leq n} x_i'^k . x_i'^l \qquad (2.13)$$

In fact, it is easy to see this result from the following relation:

$$\sum_{1 \leq i \leq n} (x_i^k - \eta_k) . (x_i^l - \eta_l) = s(k, l) - n . \eta_k . \eta_l \qquad (2.14)$$

Hence, $S(k, l)$ corresponds to the Euclidean scalar product between the kth and the lth column vectors of the transformed data table (see (2.12)).

In the case where η_k is constant for all k ($1 \leq k \leq p$) (condition C_1), the similarity coefficients s and S are equivalent with respect to the associated preordonance (see Sect. 4.2.2 of Chap. 4 of [4]).

Due to the fact that S provides correction of s in the general case where η_k ($1 \leq k \leq p$) is non-constant, all of the proximity analysis (see Sect. 2.4.4 and Sect. 2.4.5) address the S data. However, in the next section (see Sect. 2.3.2), the column association coefficient S will be expressed in the case of a constant η_k value, $1 \leq k \leq p$.

2.3.3   Equation of the Association Coefficient in the Case of a σ Form

Our purpose consists here of specifying the association coefficient $S(k, l)$ (see (2.9)) between two column attributes k and l, $1 \leq k < l \leq p$, for a σ form of the incidence data table (see (2.5 and Fig. 2.1) in the case where condition C_1 is satisfied, that is to say, in the case where the object number possessing a given attribute is constant whatever is the latter. This constant number is expressed by the $e = n . \eta$. In this instance $S(k, l)$ (see (2.9)) becomes

$$S(k, l) = \frac{s(k, l) - n . \eta^2}{\eta . \sqrt{n}} \qquad (2.15)$$

The substitution (2.12) is now defined as

$$\left(\forall(i, j), 1 \leq i \leq n, 1 \leq j \leq p,\right), x_i'^j = \frac{x_i{}^j - \eta}{\sqrt{\eta} . \sqrt{n}} \longrightarrow x_i'^j \qquad (2.16)$$

and $S(k, l)$ can be expressed by the same equation as (2.13), namely,

$$S(k, l) = \sum_{1 \leq i \leq n} x_i'^k . x_i'^l \qquad (2.17)$$

The respective values of $x_i'^k . x_i'^l$ are

$$\frac{\eta}{\sqrt{n}} \text{ if } x_i^k = x_i^l = 0$$

$$\frac{-(1 - \eta)}{\sqrt{n}} \text{ if } x_i^k = 0 \text{ and } x_i^l = 1$$

$$\frac{-(1 - \eta)}{\sqrt{n}} \text{ if } x_i^k = 1 \text{ and } x_i^l = 0$$

$$\frac{(1 - \eta)^2}{\eta . \sqrt{n}} \text{ if } x_i^k = x_i^l = 1 \qquad (2.18)$$

Therefore, the expression of $S(k, l)$ (see (2.17)) can be detailed as follows.

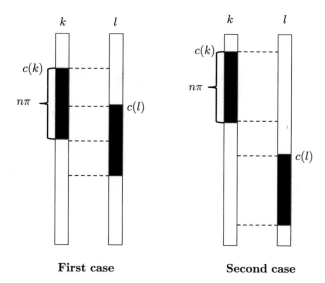

First case   **Second case**

Fig. 2.2 Two column comparison cases in a σ form

First case: $(\mathcal{O}_k \cap \mathcal{O}_l \neq \emptyset) \Leftrightarrow c(l) < c(k) + n.\eta$

$$S(k,l) = \sum_{1 \le i \le c(k)-1} \frac{\eta}{\sqrt{n}} - \sum_{c(k) \le i \le c(l)-1} \frac{1-\eta}{\sqrt{n}}$$

$$+ \sum_{c(l) \le i \le c(k)+n.\eta-1} \frac{(1-\eta)^2}{\eta.\sqrt{n}}$$

$$- \sum_{c(k)+n.\eta \le i \le c(l)+n.\eta-1} \frac{1-\eta}{\sqrt{n}} + \sum_{c(l)+n.\eta \le i \le n} \frac{\eta}{\sqrt{n}} \qquad (2.19)$$

Consider the calculation development up to the multiplicative factor $\frac{1}{\eta.\sqrt{n}}$. By grouping the first and the last terms, we obtain

$$[n(1-\eta) - (c(l) - c(k))].\eta^2$$

The second and the penultimate terms give

$$-2.[(c(l) - c(k)).\eta.(1-\eta)]$$

It remains the middle term which is equal to

$$[n.\eta - (c(l) - c(k))].(1-\eta)^2$$

The global sum is equal to

$$n(1-\eta)\eta - (c(l) - c(k))$$

And, by integrating the multiplicative factor $\frac{1}{\eta.\sqrt{n}}$, we obtain

$$S(k,l) = \sqrt{n}.(1-\eta) - \frac{1}{\eta.\sqrt{n}}.(c(l) - c(k))$$

The latter can be written as follows:

$$S(k,l) = \frac{\sqrt{n}}{\eta}[\eta.(1-\eta) - (c_l - c_k)] \qquad (2.20)$$

where

$$(\forall j), 1 \le j \le p, c_j = \frac{c(j)}{n}$$

Now, let us specify $S(k, k)$ by applying directly (2.14) in the case where $l = k$. We have

$$(x_i'^k)^2 = \begin{cases} \frac{\eta}{\sqrt{n}} & \text{if } x_i^k = 0, \\ \frac{(1-\eta)^2}{\eta \cdot \sqrt{n}} & \text{if } x_i^k = 1. \end{cases}$$

Therefore,

$$S(k, k) = \frac{\eta}{\sqrt{n}}[(n - e) + \frac{(1 - \eta)^2}{\eta^2} e]$$

After reduction, we obtain

$$S(k, k) = \sqrt{n}(1 - \eta) \tag{2.21}$$

In these conditions, the following relation is derived:

$$\frac{\eta}{\sqrt{n}}\big(S(k, k) - S(k, l)\big) = c_l - c_k$$

Second case: $(\mathcal{O}_k \cap \mathcal{O}_l) = \emptyset \Leftrightarrow c(l) \geq c(k) + n.\eta$

$$S(k, l) = \sum_{1 \leq i \leq c(k)-1} \frac{\eta}{\sqrt{n}} - \sum_{c(k) \leq i \leq c(k)+n.\eta-1} \frac{1 - \eta}{\sqrt{n}}$$

$$+ \sum_{c(k)+n.\eta \leq i < c(l)-1} \frac{\eta}{\sqrt{n}} - \sum_{c(l) \leq i \leq c(l)+n.\eta-1} \frac{1 - \eta}{\sqrt{n}}$$

$$+ \sum_{c(l)+n.\eta \leq i \leq n} \frac{\eta}{\sqrt{n}} \tag{2.22}$$

The following calculation of the second member is developed up to the multiplicative factor $\frac{1}{\sqrt{n}}$.

The first and the last terms give

$$\big(n(1 - \eta) + (c(k) - c(l))\big).\eta$$

We obtain from the second and the penultimate terms:

$$-2n\eta.(1 - \eta)$$

The middle term is equal to

$$\big((c(l) - c(k)) - n\eta\big).\eta$$

In all, we have $-n.\eta$ and then by reintroducing the multiplicative factor $\frac{1}{\sqrt{n}}$, we obtain

$$S(k, l) = -\sqrt{n}.\eta \text{ if } c(l) \geq c(k) + e$$

Finally, we get

$$S(k, l) = \frac{\sqrt{n}}{\eta}[(c_l - c_k) + \eta.(1 - \eta)] \text{ if } c(l) < c(k) + n.\eta$$

$$S(k, l) = -\sqrt{n}.\eta \text{ if } c(l) \geq c(k) + n.\eta \tag{2.23}$$

Now, it is natural to introduce a *distance* notion between column attributes as follows:

$$D(k, l) = (c_l - c_k) \text{ if } c(k) < c(l) < c(k) + e$$

$$D(k, l) = \eta \text{ if } c(l) \geq c(k) + e \tag{2.24}$$

where—recall—$\eta = \frac{e}{n}$.

In these conditions, we have the following result.

Theorem 1

$$D(k, l) = \eta.\sqrt{n}(S(k, k) - S(k, l)) \text{ for any } (k, l), 1 \leq k \leq l \leq l \leq p$$

2.3.4 *Representation on a Directed Unit Line Segment*

The incidence data table is supposed to be able to be reduced to a σ form strongly *chained vertically* (see Definition 2). In this case, there exists a column permutation followed by a row permutation which puts this specific form. In Fig. 2.2, we have to suppose that the first and the last column segments whose entries are weighted by the 1 values have a non-empty intersection. More formally, according to (2.23), we have $c(p) < c(1) + e$ and then

$$(\forall(k, l), 1 \leq k < l \leq p), S(k, l) = \frac{\sqrt{n}}{\eta}[(c_l - c_k) + \eta.(1 - \eta)] \tag{2.25}$$

In these conditions, (2.23) becomes

$$(\forall(k, l), 1 \leq k < l \leq p), D(k, l) = (c_l - c_k) \tag{2.26}$$

Therefore, we have the

Theorem 2 *If an incidence data table satisfies the following conditions:*

1. *The number of the* 1 *value component in a given column is the same for every column (for any k, $1 \leq k \leq p$, $n.\eta_k = n.\eta$);*
2. *The incidence data table is strongly vertically chained (see Definition 2);*
3. *The incidence data table can be reduced to a σ form.*

   Then there exists a representation of the data table columns by points of a directed unit segment such that the latter points are ordered according to σ and for which the mutual distances are the $D(k, l)$, $1 \leq k < l \leq p$.

   As a matter of fact, the jth column from left to right, associated with a σ form, is represented by the point of abscissa c_j of the unit interval $[0, 1]$ of the axis of reals, $1 \leq j \leq p$. Since condition 2 above is equivalent to $c_p < c_1 + \eta$, the differences $c_l - c_k$ are translated into the distances $D(k, l)$ between the columns k and l, $1 \leq k < l \leq p$.

   As expressed at the end of Sect. 2.2.2, the order of the columns for a σ form can be determined, up to a reverse permutation, from an extreme column by successive chaining. Being here in the case where the columns are mutually distinct (condition C_0), an extreme column attribute is such that there exists only a single object for which the only attribute present in it is an extreme column attribute. However, this property is too vulnerable in order to be exploited in real cases where a σ form dominates statistically but not exactly (mathematically). Precisely, the objective of the next section (see Sect. 2.4) consists of defining a *robust* statistical criterion for the determination of an extreme element.

2.4 Simultaneous Mean and Variance Analysis of Pairwise Proximities

2.4.1 Preamble

The previous extremal property (see also Theorem 2) is the starting point of an original approach of geometrical representation of the incidence data table columns. This representation enables a σ form of this table to be demonstrated. The simplest case is that for which the σ form is strongly chained. The weakly chained case will also be considered. Importantly, we introduce the case where the σ form is itself constituted by homogeneous blocks corresponding each to a σ form. These blocks may emerge from crossing two clusterings: the first, on the column attribute set and the second, on the row object set (see [4] and more particularly Chap. 8 of the latter book). Another approach for block seriation is given in [11]. The method presented here is directly based on simultaneous mean and variance analyses of pairwise proximities between the column attributes (see Sect. 2.4.3). The general equation of proximity variance analysis is given in Sect. 2.4.4. In Sect. 2.4.5, a general method of geometrical representation of the column attributes is defined. The latter is based on the determination of the attraction poles. These play a specific role with respect to proximity variance analysis. This method will be interpreted with respect to principal component analysis in Sect. 2.4.6.

2.4.2 Optimal Properties of an Extreme Point of Rectilinear Cloud

There is no restriction in the following development if we suppose the cloud of points situated on a directed linear segment of *unit* length. The latter can be denoted by $\overrightarrow{AB}$. Now, let $(c_1, c_2, ..., c_j, ..., c_p)$ be an ordered sequence of abscissas of p ordered points (from left to right) of $\overrightarrow{AB}$:

$$c_1 < c_2 < ... < c_j < ... < c_p$$

Each of the different points will be indicated by its rank, thereby j is the point of $\overrightarrow{AB}$ whose abscissa is c_j, $1 \leq j \leq p$. We have

Proposition 3 *The point which maximizes the distance average of this point to all of the cloud points is necessarily an extreme point of the rectilinear cloud.*

In more formal words, by denoting $\mathcal{M}_j$ the distance average of j to all of the points:

$$(\forall j, 1 \leq j \leq p), \mathcal{M}_j = \frac{1}{p} \sum_{1 \leq h \leq p} d(j, h) \tag{2.27}$$

we have

$$max_{1 \leq j \leq p} \mathcal{M}_j = max\{\mathcal{M}_1, \mathcal{M}_p\} \tag{2.28}$$

To begin, let us compare $\mathcal{M}_1$ with $\mathcal{M}_k$ for $k \leq p/2$. The sequence of the distances from the element 1 is

$$c_1 - c_1 = 0, c_2 - c_1, c_3 - c_1, ..., c_k - c_1, ..., c_p - c_1 \tag{2.29}$$

Then, the distance average from 1 is

$$\mathcal{M}_1 = \frac{1}{p} \left(\sum_{1 \leq j \leq p} c_j - p.c_1 \right) \tag{2.30}$$

Now, the distance sequence from k, $1 \leq k \leq p$, is:

$$c_k - c_1, c_k - c_2, ..., c_k - c_{k-1}, c_k - c_k = 0, c_{k+1} - c_k, ..., c_p - c_k \tag{2.31}$$

Hence, the distance sum from k can be put as

$$(2k - p - 1)c_k - \sum_{1 \le j \le k-1} c_j + \sum_{k+1 \le j \le p} c_j$$

On the other hand, the content of () in the $\mathcal{M}_1$ expression (see (2.30)) can be written as follows:

$$\sum_{1 \le j \le k} c_j + \sum_{k+1 \le j \le p} c_j - p.c_1$$

Therefore,

$$\mathcal{M}_1 - \mathcal{M}_k = \frac{1}{p}\left((p - 2k + 1)(c_k - c_1) - 2(k - 1)c_1 - c_1 + 2. \sum_{1 \le j \le k-1} c_j + c_k\right)$$

Finally, we have

$$\mathcal{M}_1 - \mathcal{M}_k = \frac{1}{p}\left((p - 2k + 2).(c_k - c_1) + 2 \sum_{1 \le j \le k-1} (c_j - c_1)\right) \tag{2.32}$$

Now, since

$$p - 2k + 2 > 0,\, c_k - c_1 > 0 \text{ and } c_j - c_1 \ge 0 \text{ for } 1 \le j \le k$$

We have

$$\mathcal{M}_1 - \mathcal{M}_k > 0 \text{ for all } k \text{ such that } 1 < k \le \frac{p}{2}$$

Clearly, the latter inequality remains true if p is odd and if $2k = p + 1$. Finally,

$$\mathcal{M}_1 = max_{1 \le k \le [(p+1)/2]} \mathcal{M}_k$$

where $[(p + 1)/2]$ designates the integer part of $(p + 1)/2$.

Symmetrically, by scanning the distances from right to left, we have

$$\mathcal{M}_p = max_{k > [(p+1)/2]} \mathcal{M}_k \qquad \qquad \square$$

Following the proved property, if a distance function is given on a rectilinear cloud of points, then that point which maximizes the distance average (mean) to all of the points is necessarily an extreme point (on the left or on the right). In these conditions, up to its reverse, the order of the points can be retrieved by using—through successive chaining algorithm (see Sect. 2.3.4)—the criterion maximized. Without loss of generality, the cloud points can be supposed as distributed on an oriented unit segment $\overrightarrow{AB}$.

Proposition 4 *The point of a rectilinear cloud of points which maximizes the distance variance to all of the points is necessarily an extreme point of the rectilinear cloud (on the left or on the right).*

Let us compare the distance variance from the element 1 (on the extreme left) with that from k, $1 \leq k \leq p - 1$. We have

$$V(1) = \frac{1}{p} \sum_{1 \leq j \leq p} (c_j - c_1)^2 - \frac{1}{p^2} \left(\sum_{1 \leq j \leq p} c_j - p.c_1 \right)^2$$

$$V(k) = \frac{1}{p} \sum_{1 \leq j \leq p} (c_j - c_k)^2$$

$$- \frac{1}{p^2} \left(\sum_{1 \leq j \leq k} (c_k - c_j) + \sum_{k+1 \leq j \leq p} (c_j - c_k) \right)^2 \qquad (2.33)$$

Consider 1 as origin and introduce

$$(\forall j, 1 \leq j \leq p), d_j = c_j - c_1 \qquad (2.34)$$

then the expressions of $V(1)$ and $V(k)$ become, respectively,

$$V(1) = \frac{1}{p} \sum_{1 \leq j \leq p} d_j^2 - \left(\frac{1}{p} \sum_{1 \leq j \leq p} d_j \right)^2$$

$$V(k) = \frac{1}{p} \sum_{1 \leq j \leq p} (d_j - d_k)^2$$

$$- \frac{1}{p^2} \left(\sum_{1 \leq j \leq k} (d_k - d_j) + \sum_{k+1 \leq j \leq p} (d_j - d_k) \right)^2 \qquad (2.35)$$

By using the numerical identity $x^2 - y^2 = (x + y) \times (x - y)$, we obtain

$$V(1) - V(k) = \frac{d_k}{p} \sum_{1 \leq j \leq p} (2d_j - d_k)$$

$$- \frac{1}{p^2} \left(2 \sum_{k+1 \leq j \leq p} d_j - (p - 2k)d_k \right) \times \left(2 \sum_{1 \leq j \leq k} d_j + (p - 2k)d_k \right) \qquad (2.36)$$

Putting $e_j = \frac{d_j}{d_k}$ for all j, $1 \le j \le p$, the previous expression becomes

$$\mathcal{V}(1) - \mathcal{V}(k) = \frac{4.d_k^2}{p^2}\{\frac{p}{2}\cdot \sum_{1 \le j \le p}(e_j - \frac{1}{2})$$

$$-[\sum_{1 \le j \le k} e_j + (\frac{p}{2} - k)] \times [\sum_{k+1 \le j \le p} e_j - (\frac{p}{2} - k)]\} \qquad (2.37)$$

By introducing

$$E_k = \sum_{1 \le j \le k} e_j \text{ and } F_k = \sum_{k+1 \le j \le p} e_j$$

the expression in the round brackets can be written as follows:

$$\frac{p}{2}\cdot(E_k + F_k - \frac{p}{2}) - [E_k + (\frac{p}{2} - k)] \times [F_k - (\frac{p}{2} - k)]$$

$$= [F_k - (p - k)] \times [k - E_k]$$

We have

$$e_j = \frac{d_j}{d_k} < 1 \text{ for all } j,\, j < k \text{ and } e_k = 1$$

$$e_j = \frac{d_j}{d_k} > 1 \text{ for all } j,\, j \ge k + 1 \qquad (2.38)$$

The first of Eqs. (2.38) implies $k - E_k > 0$ and the second one, $F_k - (p - k) > 0$. Consequently,

$$\mathcal{V}(1) > max\{\mathcal{V}(k)|2 \le k \le p - 1\} \qquad (2.39)$$

In the development above, the distance variance is calculated from the extreme left point 1 according to the direction "left to right". Equally, we may consider the distances from right to left and this by taking the point p as origin. In these conditions, the distance of the point p to the point j can be written as

$$(\forall j, 1 \le j \le p), d'_j = c_p - c_j \qquad (2.40)$$

More generally, the distance distribution from the point k, counted from right to left, can be written as follows:

$$c_k - c_j = d'_j - d'_k \text{ for all } j \le k$$

$$c_j - c_k = d'_k - d'_j \text{ for all } j > k \qquad (2.41)$$

According to the proof above, by symmetry, p is the point for which the distance variance—from right to left—is maximum:

$$\mathcal{V}(p) > max\{\mathcal{V}(k)|2 \leq k \leq p - 1\} \tag{2.42}$$

Finally, the point for which the distance variance to the other points is the greatest is necessarily among the points 1 or p.

□

This extremal property is also obtained by using the statistic of absolute moment of order 2. The proof given here is very different from that proposed in [8]. The respective values of the statistic considered are

$$(\forall k, 1 \leq k \leq p), \mathcal{M}_2(k) = \frac{1}{p} \sum_{1 \leq j \leq p} (c_k - c_j)^2 \tag{2.43}$$

Proposition 5 *The point of a rectilinear cloud of points which maximizes the absolute moment of order 2 of its distances to all of the rectilinear cloud points is necessarily an extreme point of this cloud (on the left or on the right).*

Up to the multiplicative factor $\frac{1}{p}$, $\mathcal{M}_2(1)$ and $\mathcal{M}_2(k)$ can be written as follows:

$$\mathcal{M}_2(1) = \sum_{1 \leq j \leq p} (c_j - c_1)^2$$

$$\mathcal{M}_2(k) = \sum_{1 \leq j \leq p} (c_j - c_k)^2 \tag{2.44}$$

Hence,

$$\mathcal{M}_2(1) - \mathcal{M}_2(k) = \sum_{1 \leq j \leq p} [(c_j - c_1) + (c_j - c_k)](c_k - c_1) \tag{2.45}$$

In these conditions, the sign of the first member of the equation above is that of

$$\sum_{1 \leq j \leq p} [2c_j - (c_1 + c_k)]$$

which is positive if and only if

$$\frac{1}{p} \sum_{1 \leq j \leq p} c_j > \frac{1}{2}(c_1 + c_k) \tag{2.46}$$

For a given k, $1 < k < p$, the left member of the previous equation can be decomposed as

$$\varphi(k) = \frac{k-1}{p} \times \frac{1}{k-1} \sum_{1 \le j \le k-1} c_j + \frac{p-k+1}{p} \times \frac{1}{p-k+1} \sum_{k \le j \le p} c_j \quad (2.47)$$

where $\varphi(k)$ is constant with respect to the k value fixing the decomposition.
   Now, let us consider

$$\psi(k) = \frac{k-1}{p}.c_1 + \frac{p-k+1}{p}.c_k \quad (2.48)$$

Due to the inequalities

$$\frac{1}{k-1} \sum_{1 \le j \le k-1} c_j > c_1$$

and

$$\frac{1}{p-k+1} \sum_{k \le j \le p} c_j > c_k$$

We have

$$\varphi(k) > \psi(k) \quad (2.49)$$

This gives for $k-1 = \frac{p}{2}$ the inequality required in (2.46). The fact that p may be odd is immaterial.
   Notice that the ψ function is decreasing. In fact, we have for the derivative

$$\psi'(k) = \frac{1}{p}(c_1 - c_k) < 0$$

Therefore

$$\psi(1) > \psi(2) > \dots > \psi([\frac{p}{2}]) \ge \psi(\frac{p}{2}) > \psi(\frac{p}{2} + 1)$$

In these conditions,
$$\mathcal{M}_2(1) = max_{1 \le k \le [\frac{p}{2}]}\mathcal{M}_2(k) \quad (2.50)$$

Symmetrically, by considering the rectilinear cloud of points from right to left, we have
$$\mathcal{M}_2(p) = max_{[\frac{p}{2}]+1 \le k \le p}\mathcal{M}_2(k) \quad (2.51)$$

□

   Finally, each of the three criteria, *Maximizing* distance mean, variance or absolute moment of order 2, provides an extreme point (on the left or on the right) of the rectilinear cloud of points. From the latter point, up to a reverse, the order of the points can be obtained by successive chaining (see Sect. 2.3.4).

2.4.3   Discovering from an Optimal Property, a σ Form

2.4.3.1   Cases of Strongly Chained Seriation Form

The results just obtained above apply directly in the case where the rectilinear cloud of points is associated with a σ form strongly chained vertically (see Theorem 2). Theorem 3 expresses this fact. An instructive example of a weakly chained σ form will be studied below (Sect. 2.4.3.2). Besides, as already mentioned in Sect. 2.4.1 more general situations will be considered in Sect. 2.4.3.

Theorem 3 *If an incidence data table satisfies the conditions 1, 2 and 3 of Theorem 2, we have*

1. *That of the p attribute columns which realizes the maximum value of its distance average (mean) to all of the attribute columns is necessarily an extreme element for the σ form.*
2. *That of the p attribute columns which realizes the maximum value of its distance variance, to all of the attribute columns, is necessarily an extreme element for the σ form.*
3. *That of the p attribute columns which realizes the maximum value of its distance absolute moment of order 2, to all of the attribute columns, is necessarily an extreme element for the σ form.*

In the case of a net σ form, mathematically defined (see (2.5)) and strongly chained vertically, the three criteria defined above are equivalent. The same does not go in the case of a fuzzy σ form. For this, a criterion based on distance variance (see (2.33)) integrates more intimately the mutual distances between the column attributes.

In Sect. 2.3.1, we have intuitively expressed what does mean a fuzzy σ form dominating statistically. In [8], an experimental design generating fuzzy σ forms is considered. The general structure of the latter forms is a *parallelogram* (see Definition 1). Two positive real parameters p and q, such that $0 < q < p < 1$, are introduced. These define probabilities of occurring a 1 value in a given cell of a simulated incidence data table. More precisely, if the cell concerned is inside the parallelogram zone, p is the probability of a value 1 to appear in this cell and then $(1 - p)$ is the probability of the 0 value for this cell. Respectively, if the cell concerned is outside the parallelogram zone, q and $(1 - q)$ have to be substituted for p and $(1 - p)$. Clearly, p is chosen appreciably greater than q; as an example, we may take $p = 0.9$ and $q = 0.05$. In the experiments brought, several sizes $(column - number \times row - number)$ of the incidence data table were considered: 25×50, 25×250, 50×500, ... On the other hand, strongly and weakly chained σ forms were examined.

2.4.3.2 Cases of a Weakly Chained Seriation Form

The property 2 above Theorem 3 does not hold in the general case of a σ form weakly chained. To establish this fact, let us consider a parallelogram σ form inclined enough in such a way that the first interval segment comprising successive 1 values (called *weighted* segment below) is disjoint from that analogous last segment; these two segments being associated with the first and the last columns of the incidence data table, respectively (see Fig. 2.3 and more explicitly Fig. 2.4).

To begin, let us give a more precise status to the notion of a σ parallelogram form. Designating by s the common length (number of cells) of the weighted vertical segments of the different columns, we put

$$e = n \times \eta \tag{2.52}$$

where η is the proportion of rows whose meeting cell with a given column contains a 1 value (see Fig. 2.4).

By considering the equation of Definition 1, we have

$$\gamma(l) - \gamma(l - 1) = \alpha \tag{2.53}$$

for $2 \leq l \leq p$. α is a portion of η such that $n.\alpha$ is an integer. In fact, $n.\alpha$ defines the vertical translation to the bottom in passing from the $(l-1)$th column to the lth one, $1 \leq l \leq (p-1)$. Therefore, we have

$$n = \eta.n + (p-1)\alpha.n \tag{2.54}$$

Fig. 2.3 Parallelogram σ form 1 weakly chained

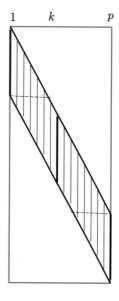

Fig. 2.4 Parallelogram σ
form 2 weakly chained

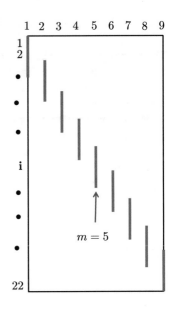

and then

$$1 = \eta + (p-1).\alpha \qquad (2.55)$$

Given the row and column numbers n and p, related by the relation (2.54), the analysis can be led with respect to a single parameter. That we will consider is $q = \frac{n}{\alpha}$. Hence, (2.55) can be written as

$$(p-1+q).\alpha = 1 \qquad (2.56)$$

From this relation, we deduce the following:

$$q = (\frac{\eta}{1-\eta}).(p-1) \qquad (2.57)$$

For simplicity reasons, we suppose p an odd integer number and we write it as

$$p = 2r+1$$

where r is an integer number. In these conditions, the rank designated by m of the median column is equal to $r+1$ (see Fig. 2.4).

In the following, two distance distributions are considered. The first one (resp., the second one) denoted by $\mathcal{D}(1)$ (resp., $\mathcal{D}(m)$) is defined by the respective distances of the column 1 (resp., the column m) to the other columns. The objective, here, consists of comparing the means, absolute moments of order 2 and variances of both distributions. In the following m is defined by the median position.

Now, let us make explicit the distributions $\mathcal{D}(1)$ and $\mathcal{D}(m)$. According to Definition 1, a σ parallelogram form weakly chained is expressed by

$$D(k, l) = \begin{cases} c_l - c_k = \alpha.(l - k) & \text{if } c_l \leq c_k + \eta \\ \eta & \text{if } c_l \geq c_k + \eta \end{cases}$$

where $1 \leq k \leq l \leq p$.

In these conditions

$$\alpha, 2\alpha, \ldots, (q-1)\alpha, \eta, \eta, \ldots, \eta \tag{2.58}$$

defines $\mathcal{D}(1)$, that is, the sequence of the distance values from the 1 column to the columns 2, 3, 4, ..., p. This sequence ends with $p - q = 2r - q + 1$ identical terms, all equal to η.

As for the former distance distribution, $\mathcal{D}(m)$ is defined by the sequence

$$\alpha, 2\alpha, \ldots, (q-1)\alpha, \eta, \eta, \ldots, \eta, \alpha, 2\alpha, \ldots, (q-1)\alpha, \eta, \eta, \ldots, \eta \tag{2.59}$$

where the sequence of the ordered first values is defined by the respective distances of m to $(m-1)$, $m-2$, ..., 2, 1. On the other hand, the respective distances of m to $(m+1)$, $(m+2)$, ..., $(p-1)$, p are given in the continuation of this sequence. Notice that in this statement a given column is identified by its label.

$(r - q + 1)$ terms equal to η end the first half (resp., the second half) of the entire sequence in (2.59).

Let $\mathcal{M}(j)$, $\mathcal{M}_2(j)$ and $\mathcal{V}(j)$ designate the mean, absolute moment of order 2 and variance of the following distance distribution

$$\mathcal{D}(j) = \{D(j, l)|l \in \{1, 2, \ldots, k, \ldots, p\} - \{j\}\} \tag{2.60}$$

Now, we shall compare $\mathcal{M}(1)$ with $\mathcal{M}(m)$, $\mathcal{M}_2(1)$ with $\mathcal{M}_2(m)$ and $\mathcal{V}(1)$ with $\mathcal{V}(m)$.

Comparison of $\mathcal{M}(1)$ with $\mathcal{M}(m)$

$$\mathcal{M}(1) = \frac{1}{2r}[\frac{(q-1)q}{2}.\alpha + (2r - q + 1).\eta] = \frac{1}{4r}[(q-1)q.\alpha + 2(2r - q + 1).\eta] \tag{2.61}$$

where $2r = p - 1$.

$$\mathcal{M}(m) = \frac{1}{2r}[(q-1)q.\alpha + 2(r - q + 1).\eta] = \frac{1}{4r}[2(q-1)q.\alpha + 4(r - q + 1).\eta] \tag{2.62}$$

Due to $q.\alpha = \eta$, we obtain

$$\mathcal{M}(1) - \mathcal{M}(m) = \frac{1}{4r}(q - 1).\eta \tag{2.63}$$

Consequently,

$$\mathcal{M}(1) > \mathcal{M}(m) \tag{2.64}$$

Comparison of $\mathcal{M}_2(1)$ **with** $\mathcal{M}_2(m)$ By taking into account the formula giving the sum of squares of the first integer numbers from 1 to $(q-1)$, we have, by considering (2.58)

$$\mathcal{M}_2(1) = \frac{1}{2r}[\frac{(q-1)q(2q-1)}{6}.\alpha^2 + (p-q).\eta^2] \tag{2.65}$$

On the other hand, by considering (2.59), we have

$$\begin{aligned}\mathcal{M}_2(m) &= \frac{1}{2r}[2\frac{(q-1)q(2q-1)}{6}.\alpha^2 + 2(r-(q-1)).\eta^2] \\ &= \frac{1}{2r}[\frac{(q-1)q(2q-1)}{3}.\alpha^2 + ((p-1)-2(q-1)).\eta^2]\end{aligned} \tag{2.66}$$

Up to the multiplicative factor $\frac{1}{2r}$, where $2r = p-1$, we have

$$\mathcal{M}_2(1) - \mathcal{M}_2(m) = -\frac{(q-1)q(2q-1)}{6}.\alpha^2 + [((p-q)-2(\frac{(p-1)}{2}-(q-1)))].\eta^2 \tag{2.67}$$

This gives

$$\mathcal{M}_2(1) - \mathcal{M}_2(m) = (q-1)[\eta^2 - \frac{q(2q-1)}{6}.\alpha^2] \tag{2.68}$$

By considering $\alpha = \frac{\eta}{q}$, we obtain

$$\mathcal{M}_2(1) - \mathcal{M}_2(m) = (q-1)[1 - \frac{(2q-1)}{6q}].\eta^2 \tag{2.69}$$

Now

$$1 > \frac{(2q-1)}{6q}$$

Therefore,

$$\mathcal{M}_2(1) > \mathcal{M}_2(m) \tag{2.70}$$

Comparison of $\mathcal{V}(1)$ **with** $\mathcal{V}(m)$ According to (2.61) and $q.\alpha = \eta$, $\mathcal{M}(1)$ can be written as follows:

$$\mathcal{M}(1) = \frac{\eta}{4r}.[4r - (q-1)] = \eta.[1 - \frac{(q-1)}{4r}] \tag{2.71}$$

It follows that

$$V(1) = M_2(1) - M(1)^2 = \eta^2 \cdot \{\frac{1}{2r} \cdot [\frac{(q-1)(2q-1)}{6q} + (2r+1) - q] - [1 - \frac{(q-1)}{4r}]^2\}$$

(2.72)

On the other hand, according to (2.62) and $q.\alpha = \eta$, we obtain

$$M(m) = \frac{\eta}{2r} \cdot (2r - q + 1)$$

(2.73)

Consequently, due to (2.66) we have

$$V(m) = \eta^2 \cdot \{\frac{1}{2r} [\frac{(q-1)(2q-1)}{3q} + 2(r - q + 1)] - \frac{1}{4r^2}(2r - q + 1)^2\} \quad (2.74)$$

Now, up to the multiplicative factor η^2, the difference $V(m) - V(1)$ can be written as follows:

$$\frac{1}{2r}[\frac{(q-1)(2q-1)}{6q} - (q-1)] - \frac{1}{4r^2}(2r - q + 1)^2 + \frac{1}{16r^2}(4r - q + 1)^2$$

(2.75)

which can be written as

$$-\frac{1}{2r}(\frac{(4q^2 - 3q - 1)}{6q}) + \frac{1}{16r^2}(8r - 3q + 3)(q - 1)$$

whose sign is that of

$$(8r - 3q + 3)(q - 1) - 8r(\frac{(4q^2 - 3q - 1)}{6q})$$

that is of

$$3[8r(q - 1)q - 3(q^2 - 1)q] - 4r(4q^2 - 3q - 1)$$

that is of

$$4r(2q^2 - 3q + 1) - 9(q^2 - 1)q$$

By considering (2.57) for which

$$r = \frac{1 - \eta}{\eta}q$$

the preceding expression becomes

$$2\frac{1-\eta}{\eta}q(2q^2 - 3q + 1) - 9(q^2 - 1)q$$

whose sign is that of

$$(4.\lambda - 9)q^2 - 6.\lambda + 9 \qquad (2.76)$$

where we have put

$$\lambda = \frac{1-\eta}{\eta}$$

The discriminant of the latter trinomial takes the following form:

$$\Delta' = 9[(\lambda - 2)^2 + 5] \qquad (2.77)$$

Therefore, the respective roots are

$$q' = \frac{3.\lambda - \sqrt{9(\lambda - 2)^2 + 5}}{4.\lambda - 9}$$

$$q'' = \frac{3.\lambda + \sqrt{9(\lambda - 2)^2 + 5}}{4\lambda - 9} \qquad (2.78)$$

The proportion η is expected to be small enough so that $4\lambda - 9 > 0$; the latter inequality corresponding to $\eta < \frac{4}{13}$. As illustrative examples: for $\eta = 0.25$, $4\lambda - 9 = 3$ and for $\eta = 0.1$, $4\lambda - 9 = 27$. Notice that the function λ of η is strictly decreasing. For $\eta = 0.25$, we obtain

$$q' = \frac{9 - \sqrt{9 + 5}}{3} \sim 1.75$$

and

$$q'' = \frac{9 + \sqrt{9 + 5}}{3} \sim 4.25$$

Hence, $V(m) > V(1)$, for $q = 1$ or for $q \geq 5$, and $V(1) > V(m)$ for $2 \leq q \leq 5$. Now, for $\eta = 0.1$, we have $\lambda = 9$. It follows that

$$q' = \frac{27 - \sqrt{9 \times 49 + 5}}{27} \sim 0.22$$

and

$$q'' = \frac{27 + \sqrt{9 \times 49 + 5}}{27} \sim 1.78$$

Hence, $\mathcal{V}(m) > \mathcal{V}(1)$, for $q \geq 2$ and $\mathcal{V}(1) > \mathcal{V}(m)$ for $q = 15$.
Finally, we have proved the following theorem:

Theorem 4 *For a σ parallelogram form weakly chained, we have*

$$\mathcal{M}(1) > \mathcal{M}(m)$$
$$\mathcal{M}_2(1) > \mathcal{M}_2(m) \tag{2.79}$$

On the other hand, we have

$$\mathcal{V}(1) > \mathcal{V}(m) \text{ for } q' < q < q''$$
$$\text{and}$$
$$\mathcal{V}(m) > \mathcal{V}(1) \text{ for } q < q' \text{ or } q > q'' \tag{2.80}$$

where $\mathcal{M}(j)$ (mean), $\mathcal{M}_2(j)$ (absolute moment of order 2) and $\mathcal{V}(j)$ (variance) are parameters already introduced above relatively to the distribution $\mathcal{D}(j)$ of the distances to the j column of the different columns other than the jth one, $1 \leq j \leq p$.
* q' and q'' are defined in (2.78) where we suppose the condition $\eta < \frac{4}{13}$, which is very generally satisfied.*

2.4.4   Block Seriation from Simultaneous Mean and Variance Analysis of Pairwise Proximities; The "Attraction Poles" Method

2.4.4.1   Preamble

We shall examine in this section mathematical examples for which the σ form is constituted of more than a single parallelogram block. In these, two consecutive parallelogram blocks are disjoint or such that the last column of the first one and the first column of the second one have at most in common a small interval of the row set equally weighted by a sequence of "one" values (see Figs. 2.5 and 2.6).

These examples in which η_k is constant ($\eta_k = \eta$ for all k, $1 \leq k \leq p$) will guide our intuition for specifying a geometrical representation method organizing the column attributes associated with an incidence data table. This method is able to reveal seriation phenomena.

2.4.4.2   Examples

Example 1

We consider here a particular case for which the seriation form σ is constituted by two disjoint parallelograms as shown in Fig. 2.5. In this diagram, an even value of p is taken ($p = 10$). The first column half part (from 1 to 5) determines the first block. The column half (from 6 to 10) is concerned by the second block. For each of both blocks Eqs (2.53) to (2.56) hold. In the latter p has be to be replaced by $p_1 = \frac{p}{2}$ and by $p_2 = \frac{p}{2}$ in the first and second blocks, respectively.

For each of the two blocks of this illustrative example $\eta = \frac{3}{22}$ and $\alpha = \frac{1}{11}$. The second block which starts at the 6th column is deduced by translating the first block. This translation assumes a down moving of $n\eta = 3$. Thereby, the form of the second block is identical to the first one. In addition, it is disjoint from the first block.

In fact, for the data structure described here it is interesting first, to recognize the two column clusters $\{1, 2, 3, 4, 5\}$ and $\{6, 7, 8, 9, 10\}$ and next, to reorder the elements of each class. Clearly, a clustering method enables the two blocks to be detected. Then, the matter consists of applying a seriation method to each of both blocks. However, in the approach developed here—which introduces a geometrical representation method (see Sect. 2.4.5)—the ordered blocks are determined in one step directly. These are driven by two specific elements belonging each to one of the blocks, respectively. Each of these two elements will be revealed from statistical characteristics associated with its distance distribution.

In the seriational configuration as given in Fig. 2.5 two distance distributions intervene. The first of them is the distance distribution of an extremal element of one of the blocks. This element may be 1, 5, 6, or 10. In this case, the distance distribution is

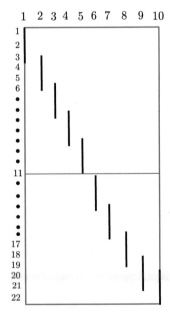

Fig. 2.5 Two disjoint σ parallelogram forms

$$\mathcal{D}_1^1 = (4, 6, 6, 6, 6, 6, 6, 6, 6)$$

The second distribution $\mathcal{D}_2^1$ is defined with respect to one of the elements 2, 3, 4, 7, 8 or 9. We have

$$\mathcal{D}_2^1 = (4, 4, 6, 6, 6, 6, 6, 6, 6)$$

Let designate by $\mathcal{M}^1(1)$, $\mathcal{M}_2^1(1)$ and $\mathcal{V}^1(1)$ the mean, absolute moment of order 2 and variance of the distance distribution $\mathcal{D}_1^1$. Similarly, $\mathcal{M}^1(2)$, $\mathcal{M}_2^1(2)$ and $\mathcal{V}^1(2)$ will designate the mean, absolute moment of order 2 and variance of the distribution $\mathcal{D}_2^1$.

$$\mathcal{M}^1(1) = \frac{52}{9} , \mathcal{M}_2^1(1) = \frac{304}{9} \text{ and } \mathcal{V}^1(1) = \frac{32}{81},$$

Besides

$$\mathcal{M}^1(2) = \frac{50}{9} , \mathcal{M}_2^1(2) = \frac{284}{9} \text{ and } \mathcal{V}^1(2) = \frac{56}{81},$$

In relation to the comparison between $\mathcal{D}_1^1$ and $\mathcal{D}_2^1$, we may notice that the distance variance with respect to an extremal element (1, 5, 6 or 10) is smaller than that with respect to an internal element (2, for example). This is due to the property of weak chaining of the σ form defined at the level of the entire incidence data table. On the contrary, the maximum value of the absolute moment of order 2 is obtained for an extremal element.

Basically, we may consider the successive chaining algorithm in order to discover from an extremal element (1, or 5, 6 or 7) the ordered sequence of the elements of a given block. Let us recall that at a given step of this algorithm, the element to pull out is the nearest of the last one extracted.

In these conditions, consider an ordered pair (k, l) of columns $(k < l)$ realizing the maximal value of the absolute moment of order 2 and such that the distance between k and l is maximum. We have

$$\mathcal{M}_2^1(k) = \mathcal{M}_2^1(l) = max\{\mathcal{M}_2(m)|1 \le m \le p\} \tag{2.81}$$

If k and l (for example, $k = 1$ and $l = 5$), the seriation obtained is restricted to the block concerned. In this case, two opposite orderings are obtained from k and l, respectively. In these conditions, one of both components k and l is omitted (say l to fix ideas) and replaced by a component l'—$l' \ne l'$—realizing the right member of (2.81). Necessarily, l' belongs to the block to which k does not belong. Equally, necessarily, the distance between k and l' is maximum. For example, if $k = 1$, l' may be 6 or 10. By this way, l' drives by successive chainings the block to which it belongs. Thus, up to a reversion, the seriation on the latter block is obtained.

Example 2

In the second configuration provided by Fig. 2.6, the second parallelogram block is deduced from the first one by applying a translation whose vector is $\overrightarrow{(1, 1)(1, 6)}$. This translation assumes a down moving of $\frac{n\eta}{3} = 1$ of the first block. In this fashion, the 10 and 11 row objects have in common the 5th and the 6th column attributes. Now, there is a single seriational block weakly chained.

The distance distributions of the column attributes to 1 and 10 columns are identical. This common distribution is

$$\mathcal{D}_1^2 = (4, 6, 6, 6, 6, 6, 6, 6, 6)$$

The distance distributions to columns 2, 3, 4, 7, 8, et 9 are identical. The common distribution is

$$\mathcal{D}_2^2 = (4, 4, 6, 6, 6, 6, 6, 6, 6)$$

The respective distance distributions of the column attributes to columns 5 and 6 are identical. The common distribution is

$$\mathcal{D}_5^2 = (2, 4, 6, 6, 6, 6, 6, 6, 6)$$

We have just above calculated statistical parameters of the distributions $\mathcal{D}_1^2$ and $\mathcal{D}_2^2$. Let us recall that we have

Fig. 2.6 Two joint σ parallelogram forms

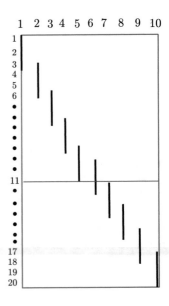

$$\mathcal{M}^2(1) = \frac{52}{9} \ , \ \mathcal{M}_2^2(1) = \frac{304}{9} \ \text{and} \ \mathcal{V}^2(1) = \frac{32}{81}$$

On the other hand,

$$\mathcal{M}^2(2) = \frac{50}{9} \ , \ \mathcal{M}_2^2(2) = \frac{284}{9} \ \text{and} \ \mathcal{V}^2(2) = \frac{56}{81}$$

Now, concerning the distance distribution $\mathcal{D}_5^2$ to the element 5, we have

$$\mathcal{M}^2(5) = \frac{48}{9} \ , \ \mathcal{M}_2^2(5) = \frac{272}{9} \ \text{and} \ \mathcal{V}^2(5) = \frac{144}{81}$$

The distance distribution which realizes the maximal value of the absolute moment of order 2 is $\mathcal{D}_1^2$:

$$\mathcal{M}_2^2(1) > max\{\mathcal{M}_2^2(2), \mathcal{M}_2^2(5)\}$$

The distance distribution which realizes the maximal value of the variance is $\mathcal{D}_5^2$:

$$\mathcal{V}^2(5) > max\{\mathcal{V}^2(1), \mathcal{V}^2(2)\}$$

In relation to the columns 1 and 2, 5 participates more intimately to the seriation than 1 or 2. Consequently, 5 is more discriminant than 1 or 2 for pulling the seriation.

Starting with the element 5, the successive chaining algorithm produces the sequence $(5, 6, 7, 8, 9, 10)$. The element 6 has the same properties as those of 5 (maximal value of the variance of its distances to the other elements). By successive chaining 6 pulls the sequence $(6, 5, 4, 3, 2, 1)$. In these conditions, the elements 5 and 6 enable together the entire seriation to be built.

2.4.5 Proximity Variance Analysis Equation

Let us return to the general expression of the data table as given in (2.2) and to the association coefficient $S(k, l)$ between two column attributes defined in (2.9), $1 \leq k < l \leq p$. The latter—with the notations of Sect. 2.3.2—is written as follows:

$$S(k, l) = \frac{s(k, l) - n.\eta_k.\eta_l}{\sqrt{n.\eta_k.\eta_l}} \tag{2.82}$$

We shall now consider a global variance analysis of the proximity table

$$\{S(k, l) | 1 \leq k \neq l \leq p\} \tag{2.83}$$

In the following, the notations $S(k, l)$ and S_{kl} are interchangeable, $1 \leq k, l \leq p$. We set

$$\bar{S}_k = \frac{1}{p-1} \sum_{\{l|1\leq l \neq k \leq p\}} S_{kl} \tag{2.84}$$

$\bar{S}_k$ defines the mean (average) of the proximities to k, $1 \leq k \leq p$.
Thereby, the global mean of the proximities between two distinct elements is

$$\bar{S} = \frac{1}{p} \sum_{k|1\leq k \leq p} \bar{S}_k \tag{2.85}$$

Let us now decompose the difference $(S_{kl} - \bar{S})$ as follows:

$$S_{kl} - \bar{S} = (S_{kl} - \bar{S}_k) + (\bar{S}_k - \bar{S})$$

Then we have

$$(S_{kl} - \bar{S})^2 = (S_{kl} - \bar{S}_k)^2 + (\bar{S}_k - \bar{S})^2 + 2(S_{kl} - \bar{S}_k).(\bar{S}_k - \bar{S})$$

By summing with respect to l and for $l \neq k$, we obtain

$$\sum_{\{l|1\leq l \neq k \leq p\}} (S_{kl} - \bar{S})^2 = \sum_{\{l|1\leq l \neq k \leq p\}} (S_{kl} - \bar{S}_k)^2 + (p-1)(\bar{S}_k - \bar{S})^2 + 0 \tag{2.86}$$

Now, by summing with respect to k and dividing the two members of the preceding equation by $p(p-1)$, we get

$$\frac{1}{p(p-1)} \sum_{\{1\leq l \neq k \leq p\}} (S_{kl} - \bar{S})^2 =$$

$$\frac{1}{p} \sum_{\{1\leq k \leq p\}} \frac{1}{p-1} \sum_{\{l|1\leq l \neq k \leq p\}} (S_{kl} - \bar{S}_k)^2$$

$$+ \frac{1}{p} \sum_{1\leq k \leq p} (\bar{S}_k - \bar{S})^2 \tag{2.87}$$

Similarly, as in (2.16) set

$$x_i'^j = \frac{x_i^j - \eta_j}{\sqrt{\eta_j.\sqrt{n}}}$$

where

$$\eta_j = \frac{1}{n} \sum_{1 \le i \le n} x_i^j$$

is nothing else than the proportion of objects for which the Boolean attribute a^j is $TRUE$, $1 \le j \le p$.

We may notice that the association coefficient $S(k, l)$ can be written as a function of the following distance index:

$$D^2(k, l) = \sum_{1 \le i \le n} (x_i'^k - x_i'^l)^2 \qquad (2.88)$$

by using the formula

$$S(k, l) = \frac{1}{2}\{-D^2(k, l) + \sqrt{n}[(1 - \eta_k) + (1 - \eta_l)]\} \qquad (2.89)$$

In fact, we have (see (2.16))

$$\sum_{1 \le i \le n} (x_i'^j)^2 = \sqrt{n}(1 - \eta_j)$$

In (2.87), the left member depends on the dispersion of the elements representing the set of the p Boolean attributes in the simplex $\mathcal{P}(\mathcal{O})$ defined by the subset of $\mathcal{O}$ (see Sects. 3.2.1 and 5.2.1 of [4]). This first member, defined by the total variance of the association coefficients $S(k, l)$ ($1 \le k \ne l \le p$), is decomposed into two parts. The first one is the mean (average) of the respective variances of the association coefficients S of each of the Boolean attributes with the other ones. This variance mean defines the *Within* column similarities variances. The second part is the *Between* column similarities variance. The latter is a measure of the distance between the similarity structure observed and that corresponding to a spherical one defined by $\bar{S}_k$ constant for every k, $1 \le k \le p$.

Testing hypotheses

These tests refer to the hypothesis $\mathcal{H}_3$ of no relation (or mutual independence) between the different Boolean attributes $a^1, a^2, ..., a^k, ..., a^p$ (see Sect. 5.2.1 of [4]). We may consider two different versions for the alternative hypothesis to the mutual independence between the Boolean attributes:

1. The alternative hypothesis is defined by a *Seriation*: the incidence data table can be reduced—up to a "small" number of cells—to a σ form;
2. The alternative hypothesis is defined by a spherical arrangement of the attribute data according to the similarity index $S(k, l)$.

The former test will be based on the statistic

$$max_{k|1\leq k\leq p}\Big(\sum_{\{l|1\leq l\neq k\leq p\}} (S_{kl} - \bar{S}_k)^2 \Big) \tag{2.90}$$

If p is not too large, these testing hypotheses can be carried out from simulating—a sufficient number of times—the incidence data table under the random hypothesis $\mathcal{H}_3$.

Otherwise, we can notice that under the $\mathcal{H}_3$ hypothesis and for $n = card(\mathcal{O})$ large enough, the sequence

$$S(k, 1), S(k, 2), ..., S(k, k-1), S(k, k+1), ..., S(k, p) \tag{2.91}$$

of the association coefficient values between a fixed column k and the other ones can be considered—under the $\mathcal{H}_3$ hypothesis—as a sequence of $(p-1)$ independent instances of the standard normal distribution. In these conditions,

$$\sum_{\{l|1\leq l\neq k\leq p\}} (S_{kl} - \bar{S}_k)^2 \tag{2.92}$$

is—for a fixed k—a realization of a χ^2 statistic with $(p-2)$ degrees of freedom. Now, notice that the p values of the latter statistic obtained, respectively, for k varying from 1 to p are not completely independent. In fact, for the p sequences as (2.91), each $S(k, l)$—for $l \neq k$—occurs in exactly two different sequences. Nevertheless, the degree of this dependence is weak. And, it is as weak as p is large. Consequently, in order to evaluate (2.90), we will refer to the distribution of the maximum value of p independent χ^2 Statistics, each with $(p-2)$ degrees of freedom.

Similarly, we will refer to the χ^2 probability law with $(p-1)$ degrees of freedom, in order to evaluate the smallness of

$$\sum_{l|1\leq k\leq p} (\bar{S}_k - \bar{S})^2 \tag{2.93}$$

Only extensive experiment enables the testing hypotheses 1 and 2 above to be provided by practical interest. In fact, several structures of statistical link, not necessarily related to a *seriation* or a *spherical arrangement*, might be shown by a non-negligible elevation (resp., diminution) of (2.90) (resp., 2.93). In general, most of these statistical link structures are compatible with a cluster structuration. Therefore, the interest of the *classifiability* test is well founded (see Sect. 8.7 of [4]).

2.4.6 Geometrical Representation by the Attraction Poles Method

2.4.6.1 Preamble

Let E designate the set to be represented. In this section, we will be concerned by the case where E is a set of descriptive Boolean attributes, these labeling the columns of an incidence data table. The basic idea in this approach is to consider for a given element e ($e \in E$) the discrimination degree of its relationships with the other elements of E. Then, we have to determine a set of a few specific elements (the *Attraction Poles*), mutually as different as possible, and to arrange the whole set E with respect to these elements. This arrangement depends on the synthetic data representation proposed. In this section, we will be interested in geometrical—mainly planar—representation. The latter will be discussed with respect to that given by component factorial analysis methods. In Chap. 6, we will define and study clustering methods based on the determination of attraction poles.

The method developed here focuses on, first, the representation of the set of the Boolean attributes indexing the columns of the incidence data table. Two remoteness notions between two elements of E can be envisaged. The first one is related to *statistical independence* notion and the second one to *metrical distance*. Clearly, these notions can be situated analytically relative to each other. However, statistical independence notion can be more associated with geometrical representation and metrical distance notion, with clustering methods.

A final but important point. As in the case of the *LLA* method a *Similarity Global Reduction* can be appropriately applied on the association coefficients $S(k, l)$, $1 \le k < l \le p$ (see (5.2.51) and (5.2.52) of Sect. 5.2.15 of Chap. 5 of [4]). For this, $S_g(k, l)$ is substituted for $S(k, l)$, according to the following formula:

$$S_g(k, l) = \frac{S(k, l) - mean_e(S)}{\sqrt{var_e(S)}}$$

where $mean_e(S)$ and $var_e(S)$ are the mean and variance of the similarity distribution $\{S(k, l)|1 \le k < l \le p\}$.

Notice that with this substitution the left member of (2.87) becomes equal to unity.

2.4.6.2 The Attraction Poles; Definition and Construction

As stated in the previous preamble, we are concerned by geometrical representation of a set $\mathcal{A}$ of Boolean attributes indexing the columns of an incidence data table $\mathcal{T}$ (see (2.1) and (2.2)). The axes of this representation result from the determination of attraction poles. The latter are built step by step as follows.

The first attraction pole P_1 is defined as the column element maximizing the *variance* of its proximities with the other column elements, namely,

$$P_1 = Arg\{max\big(\mathcal{V}(k)\big)|1 \le k \le p\} \tag{2.94}$$

where

$$\mathcal{V}(k) = \frac{1}{p-1} \sum_{\{l|1 \le l \ne k \le p\}} (S_{lk} - \overline{S_l})^2$$

In practice, very generally, there is a single element which realizes the maximum value of $\mathcal{V}$. If not, P_1 can be taken randomly as one of the elements maximizing $\mathcal{V}(k)$. P_1 will define in the geometrical representation plane the first axis. This is taken as an horizontal axis, oriented from left to right.

Now, let us consider the determination of the second attraction pole P_2. Two requirements have to be expected for P_2:

1. The proximity variance of P_2 ($P_2 \ne P_1$) with respect to the other elements is as large as possible;
2. P_2 is as independent as possible of P_1.

In these conditions, the determination criterion of P_2 has to combine

1. $\mathcal{V}(P_2)$ is as large as possible;
2. The absolute value of $S(P_1, P_2)$ is as small as possible.

In the following, P_j ($j = 1, 2, ...$) will designate the column number corresponding to the attribute which defines the jth attraction pole. Without risk of ambiguity, P_j will also indicate the latter attribute. Moreover, the point translating P_j in the geometrical representation space will also be pointed as P_j.

Now, it is natural to require from the criterion to establish, the same analytical nature as that used for detecting the first pole, that is to say, a square of a similarity S. Therefore, P_2 can be determined as follows:

$$P_2 = Arg\{max[(\frac{\mathcal{V}(k)}{S(k, P_1)})^2| \, |S(k, P_1)| \ge \epsilon, 1 \le k \ne P_1 \le p]\} \tag{2.95}$$

where the integer k indicates the kth column attribute, where $| \bullet |$ means the absolute value of $\bullet$ and where ϵ is a strictly positive real as small as possible. In fact, a zero value of $S(k, P_1)$ makes this criterion not defined.

Mostly, in real data, the condition

$$min\{|S(k, P_1)||1 \le k \ne P_1 \le p\} \ge \epsilon \tag{2.96}$$

where ϵ is a strictly real positive is satisfied.

Generally, the case where the left member of (2.96) is null results from a mathematical construction of the incidence data table. Anyway, in the latter case, a correction of the similarities

$$\{S(k, P_1)|1 \le k \ne P_1 \le p\}$$

can be proposed in such a way that condition (2.96) is satisfied (see below).

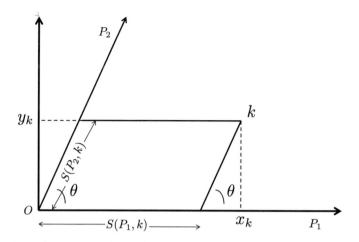

Fig. 2.7 The diagram of the representation axes

P_2 will drive the second representation axis.

The origin of the planar representation is the intersection of the two axes driven by P_1 and P_2, respectively. The latter point can be denoted by O and we may consider that its both coordinates are equal to $S(P_1, P_2)$. The angle of the axis $\overrightarrow{OP_1}$ and $\overrightarrow{OP_2}$, $\theta(\overrightarrow{OP_1}, \overrightarrow{OP_2})$ is counted in the counterclockwise direction (see Fig. 2.7). It is set as

$$\theta(\overrightarrow{OP_1}, \overrightarrow{OP_2}) = \pi.\Phi[S(P_1, P_2)] \tag{2.97}$$

where π is the angle in radians equivalent to an angle of 180 degrees and where Φ is the standard normal cumulative distribution function. Thereby, $\theta(\overrightarrow{OP_1}, \overrightarrow{OP_2})$ is as near $\pi/2$ (90 degrees) as $S(P_1, P_2)$ is near 0.

To simplify the notation put $\theta = \theta(\overrightarrow{OP_1}, \overrightarrow{OP_2})$. θ is lower (resp., greater) than $\pi/2$ if $S(P_1, P_2)$ is negative (resp., positive).

The abscissa and the ordinate of the point representing the column attribute k in the geometrical plane $P_1 O P_2$ are $S(P_1, k)$ and $S(P_2, k)$, respectively. $S(P_1, k)$ [resp., $S(P_2, k)$] is counted in parallel—with the same direction—to the axis $\overrightarrow{OP_1}$ (resp., $\overrightarrow{OP_2}$) (see Fig. 2.7).

For a better visualization, instead of the oblique axes system $(\overrightarrow{OP_1}, \overrightarrow{OP_2})$ of θ angle, we can consider a rectangular axis system $(\overrightarrow{OX}, \overrightarrow{OY})$, where $\overrightarrow{OX}$ coincides with $\overrightarrow{OP_1}$ and where $\overrightarrow{OY}$ is obtained by rotating $\overrightarrow{OX}$ of $\frac{\pi}{2}$ angle in the counterclockwise direction (see Fig. 2.7). The new coordinate system of the kth element becomes

$$x_k = S(P_1, k) + S(P_2, k).cos(\theta)$$
$$y_k = S(P_2, k).sin(\theta) \tag{2.98}$$

Now, let us deal with the case where condition (2.96) does not hold, that is to say,

$$min\{|S(P_1, k)||1 \leq k \neq P_1 \leq p\} = 0 \tag{2.99}$$

Let k_0 be a column attribute such that

$$S(P_1, k_0) = 0 \tag{2.100}$$

The left member of (2.100) can be written as

$$S(P_1, k_0) = \|P_1\| . \|k_0\| . cos(\gamma_0) \tag{2.101}$$

where $\gamma_0 = \frac{\pi}{2}$ and where $\|h\|$ $(1 \leq h \leq p)$ is the ordinary Euclidean norm of the hth vector column of which the ith component is

$$x_i^{\prime h} = \frac{x_i^{\,h} - \eta_h}{\sqrt{\eta_h} . \sqrt{n}} \tag{2.102}$$

$1 \leq i \leq n$, (see (2.12)).
As in (2.101), express $S(P_1, k)$ as follows:

$$S(P_1, k) = \|P_1\| . \|k\| . cos(\gamma_k) \tag{2.103}$$

A solution which can be proposed for the case $\gamma_0 = \pi/2$ consists of substituting $\gamma_0 - \epsilon$ for γ_0, where ϵ is the smallest value of an angle such that

$$|cos(\gamma_0 - \epsilon)| \geq min\{|cos(\gamma_k)||1 \leq k \leq p, |cos(\gamma_k)| > 0\} \tag{2.104}$$

This type of correction may equally be adopted in the development below for seeking more than two poles. And then, we will no longer mention this problem.

As a matter of fact, we can determine step-by-step additional representation axes—driven by attraction poles—revealing new dimensions. Here, two rules can be envisaged for this. The first requires logical operations and the second, numerical ones. Thus, the third axis $\overrightarrow{OP_3}$ will be defined from

$$P_3 = Arg\left\{max[min\{(\frac{\mathcal{V}(k)}{S(P_1, P_k)})^2, (\frac{\mathcal{V}(k)}{S(P_2, P_k)})^2\}] \,| \right.$$
$$\left. 1 \leq k, k \neq P_1, k \neq P_2 \leq p\right\} \tag{2.105}$$

or from

$$P_3 = Arg\left\{max\{[(\frac{\mathcal{V}(k)}{S(P_1, P_k)})^2 \times (\frac{\mathcal{V}(k)}{S(P_2, P_k)})^2]^{1/2}|\right.$$
$$\left. 1 \leq k, k \neq P_1, k \neq P_2 \leq p\}\right\} \tag{2.106}$$

With these types of rules a sequence of poles can be established. Designating by $\mathcal{P}_{j-1}$ the set of the first $(j-1)$ poles:

$$\mathcal{P}_{j-1} = \{P_h | 1 \leq h \leq j-1\} \tag{2.107}$$

The jth pole is defined by

$$P_j = Arg\left\{max[min_{1 \leq h \leq (j-1)}[(\frac{\mathcal{V}(k)}{S(P_h, P_k)})^2]\right.$$
$$\left. | 1 \leq k, k \neq P_1, k \neq P_2 \leq p\right\} \tag{2.108}$$

or from

$$P_j = Arg\left\{max\{[\Pi_{1 \leq h \leq (j-1)}(\frac{\mathcal{V}(k)}{S(P_h, P_k)})^2]^{1/(j-1)}\right.$$
$$\left. | 1 \leq k, k \neq P_1, k \neq P_2 \leq p\right\} \tag{2.109}$$

For discovering a seriation, it suffices to consider the column representation in the geometrical plane $P_1 O P_2$. This highlighting concerns particularly strongly chained seriations. A particular examination must be regarded in the case of weakly chained seriations as in Figs. 2.1 and 2.6.

However, a certain robustness of the geometrical representation can be observed. As an example, consider the case of an incidence data table configuration as that considered in Fig. 2.8, where the σ form is composed of two disjoint parallelograms and where the second one is deduced from the first one by symmetry. The latter symmetry is with respect to the extreme point of the first block (the lowest and the most on the right). In this case (see Fig. 2.9), the two first poles $F1$ and $F2$ are, respectively, located each on one of each block. The first one $F1$ (resp., the second one $F2$) is in the center of the first block (resp., the second block). $F1$ drives columns of the first block and $F2$, of the second block. This configuration will be considered again in Chap. 4.

Now, let us consider the example of Fig. 2.10 where the total σ form is a result from concatenation of three σ parallelogram forms, each strongly chained on a column interval. These intervals are mutually disjoint. Thereby, each σ form is circumscribed in a subrectangle of the data table. Designate by G, H and K these successive σ forms, each defined by what it can be called a block seriation [11]. The column interval associated with G precedes that associated with H and the column interval associated with H precedes that associated with K.

Considering what we observed in the previous example (see Figs. 2.8 and 2.9), the ordering of the column set of a given block G, H or K can be obtained by applying

Fig. 2.8 Two symmetrical
sigma parallelogram forms

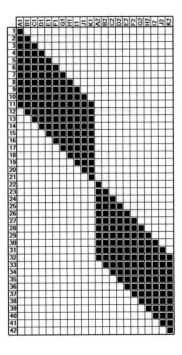

Fig. 2.9 Geometrical
representation of two disjoint
blocks seriation

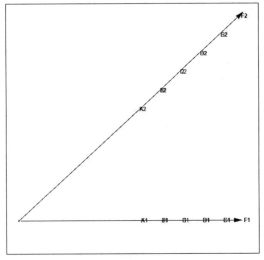

the planar geometrical representation algorithm to the subrectangle of the data table
circumscribing the block concerned. The respective boundaries of the three blocks
are easily fixed by using a clustering method on the column set of the entire incidence
data table.

Fig. 2.10 Three disjoint
sigma parallelogram forms

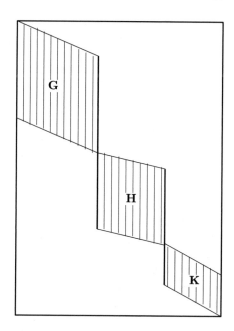

From the third pole, the determination of the sequence of poles is based on a *min-max* or a *geometrical mean* criterion (see (2.105), (2.108) and the equations which follow directly). Such a rule avoids mathematical linear calculations. Nevertheless, we may determine the sequence of poles—starting with the third one—with a criterion having the same nature as that employed to get the second pole. However, this strategy that we describe briefly in the following implies matrices inversion calculations.

By considering the coding (2.12), for two columns k and l of the incidence data table, $1 \leq k < l \leq p$, $S(l, k)$ can be interpreted—up to norm consideration—as the projection measure of the column vector k (resp., l) on the column vector l (resp. k). In relation to the geometrical space $\mathbb{R}^n$, whose origin being indicated by 0, we may denote the column vectors k and l by

$$\hat{k} = \overrightarrow{Ok} \text{ and } \hat{l} = \overrightarrow{Ol}$$

or, equivalently,

$$\hat{k} = (k - O) \text{ and } \hat{l} = (l - O)$$

the two respective column vectors, whose components are

$$(x_i'^k | 1 \leq i \leq n) \text{ and } (x_i'^l | 1 \leq i \leq n)$$

We shall note down a measure such $S(k, l)$ as $Mproj(\hat{k}, \hat{l})$. With this notation the pole P_2, corresponding to column vector $\hat{p}_2$ is defined by

$$\hat{p}_2 = Arg\{max([\mathcal{V}(j)/Mproj(\hat{j}, \hat{p}_1)]^2)|\hat{j} \neq \hat{p}_1, 1 \leq j \leq p\} \qquad (2.110)$$

Let us now consider the determination of the third pole P_3 and let $\hat{p} = (\hat{p}_1, \hat{p}_2)$ be the matrix with n rows and 2 columns corresponding to P_1 and P_2. The projector $Proj_{12}$ on the plane generated by $\hat{p} = (\hat{p}_1, \hat{p}_2) = (p_1 - O, p_2 - O)$ is defined by the matrix that we designate equally by $Proj_{12}$:

$$Proj_{12} = \hat{p}(^T\hat{p}.\hat{p})^{-1^T}\hat{p} \qquad (2.111)$$

where T indicates Transpose. Thus, the determination of $Proj_{12}$ assumes inversion of a 2×2 matrix, the rank of the latter being 2, because $(\hat{p}_2 - O)$ is not colinear with $(\hat{p}_1 - O)$.

More generally, if $\mathcal{P}_c = \{P_j|1 \leq j \leq c\}$ specifies the set of the first c poles extracted, the next pole P_{c+1} will be defined by the column vector $\hat{k}$ where $k \notin \mathcal{P}_c$ and maximizing the criterion

$$\hat{p}_{c+1} = Arg\{max[\mathcal{V}(j)/Mproj(\hat{j}, \hat{p})]^2|\hat{j} \notin \{\hat{p}_1, \hat{p}_2, ..., \hat{p}_c\}, 1 \leq j \leq p\} \qquad (2.112)$$

where $\hat{p}$ indicates the space generated by the c first poles. More formally,

$$\hat{p}_{c+1} = Arg\{max[\mathcal{V}(j)/Mproj(\hat{j}, \hat{p}_1, ..., \hat{p}_c)]^2|\hat{j} \notin \{\hat{p}_1, \hat{p}_2, ..., \hat{p}_c\}, 1 \leq j \leq p\} \qquad (2.113)$$

The projector expressed in the denominator of the preceding ratio is given by

$$P_{12...c} = \hat{p}.(^T\hat{p}.\hat{p})^{-1}.^T\hat{p} \qquad (2.114)$$

where

$$\hat{p} = (\hat{p}_1, \hat{p}_2, , ..., \hat{p}_c) \qquad (2.115)$$

$\hat{p}$ is a matrix with n rows and c columns. The rank of this matrix is expected to be c. The latter condition is generally satisfied, because, in fact, the vector columns $\hat{p}_1, \hat{p}_2, ..., \hat{p}_c$ are established in such a way to be mutually as linearly independent as possible. Therefore, inversion of a square $c \times c$ matrix is required in order to obtain $Proj_{12...c}$.

2.4.6.3   Comparing with Factorial Component Analysis

To begin let us recall how we obtain the first axis $\hat{p}_1 = P_1 - O$, driven by the first pole P_1 in the geometrical representation proposed. In fact, as seen above, the following axes are derived from the first one step by step. For this, two methods have been considered above.

As specified, $\hat{p}_1$ is defined from a column vector of the normalized version of the incidence data table (see (2.12)). More precisely, if $P_1 = k$, we have

$$\hat{p}_1 = P_1 - O =^T \left(x_i'^k = \frac{x_i^k - \eta_k}{\sqrt{\eta_{k \cdot} \sqrt{n}}} | 1 \le i \le n \right) \tag{2.116}$$

Designate here by ξ^k the column vector (2.116) with n components. According to (2.94), ξ^k maximizes

$$\sum_{\{l | 1 \le l \ne k \le p\}} \left[\xi^k \cdot \left(\xi^l - \frac{1}{p-1} \sum_{\{h | 1 \le h \ne k \le p\}} \xi^h \right) \right]^2 \tag{2.117}$$

To be convinced of this formula, consider (2.13) where the right member is the inner product between ξ^k and ξ^l, $1 \le k < l \le p$.

For the factorial component analysis method, compared with the nearest pole attraction representation method, the first axis is defined by a unit vector η, with n components maximizing

$$\sum_{1 \le l \le p} (\eta \cdot \xi^l)^2 \tag{2.118}$$

Consequently, the matter in our method—intuitively speaking—consists of mutual evaluation of the data from *inside* the data. Then, calculating eigenvalues and eigenvectors is not needed as the case is in factorial analysis and spectral methods [10].

Let us mention here that the starting point of the new representation process was the introduction of a measure of neutrality (resp., discrimination) degree of a given element with respect to the other ones of the set to be analyzed by clustering (see Sect. 8.5 of [4]). The set concerned in our development is composed of descriptive Boolean attributes.

Moreover, it appeared that in relation to the result observed of a factorial component analysis, the most neutral elements are mostly assembled around the origin and the most discriminant ones, around the extremity of the first axis, and this, especially when the explained inertia by the first factorial axis is large enough.

The geometrical representation method around attraction poles has been established here in order to plot a set of Boolean descriptive attributes. This method can be extended, naturally, to figure a set of descriptive attributes of any type. Indeed, we have seen in Chaps. 3, 4, 5 and 6 of [4], how to generalize the notion and the construction of an association coefficient between Boolean attributes to other types of attributes. In this generalization where a set theoretic representation of the descriptive attributes is emphasized, the following types are treated: *numerical, contingency, nominal categorical, ordinal categorical, ranking, binary valuation, categorical valuation* and *preordonance*.

Now, notice that the proximity variance analysis Formula (8.5.5) of Sect. 8.5 of Chap. 8 of [4] is given in the dual case concerned by the representation of an object set $\mathcal{O}$, provided by a similarity index $\mathcal{S}$. This dual case has to be studied and exploited with respect to different structures of the data table. In the next chapter, the equation mentioned will constitute a backdrop with respect to which clustering algorithms of $\mathcal{O}$, based on the determination of attraction poles, will be examined.

Let us return to the Boolean case. As expressed above (see Sect. 2.4.6), we focus on the geometrical representation of the descriptive attributes indexing the columns of the incidence data table. For a planar representation, a strongly chained parallelogram form appears as a sequence of points hugging the form of a handwritten letter **s**. This form becomes a parabolic one in the case of applying correspondence analysis method [12]. Let us indicate here that the Horse-shoe form of the Kendall method [13] concerns in fact the representation of the row set, representing the object set (a set of tombs in the archeological seriation problem).

To end this section, let us emphasize a final point already considered above. Mainly, most seriation methods are concerned first by the ordering of the row object set. As just said, the strategy we adopt begins by ordering the attribute set. Ordering the row object set is then derived by a specific algorithm that we will detail in Sect. 5.2 of Chap. 5.

References

1. M. Hahsler, K. Hornik, C. Buchta, Getting things in order: an introduction to the R package seriation. J. Stat. Softw. **25**(3), 1–34 (2008)
2. F.M.W. Petrie, Sequences in the prehistoric remains. J. Anthropol. Inst. **29**, 295–301 (1899)
3. I. Liiv, Seriation and matrix reordering methods: an overview. Wiley InterScience (www.interscience.wiley.com), (DOI: https://doi.org/10.1002), 70–91, Jan. 2010
4. I.C. Lerman, *Foundations and Methods in Combinatorial and Statistical Data Analysis and Clustering* (Springer, 2016)
5. H. Mannila, Finding total and partial orders from data for seriation, in *Discovery Science 2008, LINAI 5255*, ed. by M.R. Berthold J.-F. Boulicau, T. Horvath (Springer, 2008), pp. 16–25
6. I.C. Lerman, Analyse du phénomène de la "sériation" à partir d'un tableau d'incidence. Mathématiques et Sciences Humaines **38**, 39–57 (1972)
7. I.C. Lerman, H. Leredde, La méthode des pôles d'attraction, in *Analyse des Données et Informatique*, ed. by E. Diday et al. (IRIA, 1977), pp. 37–49
8. H. Leredde, *La méthode des poles d'attraction, La méthode des poles d'agrégation*. Ph.D. thesis, Université de Paris 6, Oct. 1979
9. I.C. Lerman, *Classification et analyse ordinale des données* (Dunod, 1981)
10. J.E. Atkins, E.G. Boman, B. Hendrickson, A spectral algorithm for seriation and the consecutive ones problem. Siam J. Comput. **28**(1), 297–310 (1998)
11. F. Marcotorchino, Block seriation problems: a unified approach. Appl. Stoch. Models Data Anal. **3**, 73–91 (1987)
12. J.-P. Benzécri, *L'analyse des données, tome II* (Dunod, 1973)
13. D.G. Kendall, Seriation from abundance matrices, in *Mathematics in Archaeological and Historical Sciences*, ed. by D.G. Kendall F.R. Hodson, P. Tautu (Chicago, Aldine-Atherton, 1971), pp. 214–252

Chapter 3
Main Approaches in Seriation: The Attraction Pole Case

3.1 Introduction

As previously, the data table crosses an object set $\mathcal{O}$ with a descriptive attribute set $\mathcal{A}$. $\mathcal{O}$ indexes the row set and $\mathcal{A}$, the column set (see (2.1) and (2.2)). The objective consists of discovering a synthesis structure defined by a couple of associated total preorders (ranking with ties) on $\mathcal{O}$ and $\mathcal{A}$, respectively. In this construction, the ranking on $\mathcal{O}$ (resp., on $\mathcal{A}$) has to respect as much as possible the similarities between the elements of $\mathcal{O}$ (resp., of $\mathcal{A}$). The rows and columns of a given data table have to be permuted in order to reveal both total preorders (rankings). In some methods, the latter (preorders) are jointly established. However, mostly the process is decomposed into two steps. First, we determine a ranking on one of the sets $\mathcal{O}$ or $\mathcal{A}$ and second, associate with it a corresponding ranking on $\mathcal{A}$ or $\mathcal{O}$, respectively. As indicated previously, we begin in our method by establishing a total preorder on $\mathcal{A}$ and then a total preorder on $\mathcal{O}$ is associated with it. Nevertheless, here, we start by describing methods of the strategy, mostly studied, in which a ranking of $\mathcal{O}$ is sought first. This, in such a way that in each column of the incidence data table, the sequence of ones determines a connected interval of the row set (elimination of the Lazarus events (see Chap. 2, Sect. 2.2.1)). For this end, an ordered pair (criterion, algorithm) is defined. The two components of the latter may be formally and even mathematically related. At each step of the algorithm progression, the criterion concerned is optimized at best. The algorithm leads to a total preorder on $\mathcal{O}$, directly or induced by a planar geometrical representation of $\mathcal{O}$. This ranking is the more so as it follows the $\mathcal{O}$ pairwise similarities. These may result from calculation or directly given. Otherwise, the criterion considered may be a local one with regard to the current algorithm status and this feature influences its computational complexity.

By denoting as in (2.1),

$$\mathcal{O} = \{o_i | 1 \le i \le n\} \tag{3.1}$$

the similarity data can be expressed as follows:

© The Author(s), under exclusive license to Springer Nature Switzerland AG 2022
I. C. Lerman and H. Leredde, *Seriation in Combinatorial and Statistical Data Analysis*, Advanced Information and Knowledge Processing,
https://doi.org/10.1007/978-3-030-92694-6_3

$$\mathcal{S} = \{S(o_i, o_j) | 1 \leq i < j \leq n\} \tag{3.2}$$

where $S(o_i, o_j)$ is the numerical similarity between the objects o_i and o_j, $1 \leq i < j \leq n$. The space complexity of $\mathcal{S}$ is $O(n^2)$.

Now, let us express here a general principle concerning data analysis algorithms. In these, there are two main types of strategies. For the first one, which we can designate by **G** (G as *Global*), each step of the algorithm optimizes the local part affected of a global criterion in which the whole table $\mathcal{S}$ intervenes. In the second type of strategy, which we can designate by **L** (L as *Local*), the algorithm optimizes a local criterion in which only a subset of $\mathcal{S}$ values is concerned.

The type **G** of approach where the goal consists of directly optimizing a global criterion may appear more satisfying conceptually. However, we have to be aware that at each algorithm step, only local optimization is obtained and, generally and mostly, the final result corresponds to a *local* optimum of the *global* criterion concerned. Additionally, the analytical form of the criterion may depend on how the target structure (seriation here) is encoded.

Differently, in a **L** approach, the optimization associated with each step of the algorithm concerns a *local* criterion. Consequently, as a result, great flexibility and fine analysis of the criterion to choose are possible. Otherwise, a global criterion, not necessarily related to the local criterion used for the construction, can be considered in order to evaluate the final result obtained. In this regard, we develop in Sect. 3.3.3 a criterion for evaluating the quality of a seriation. This criterion is based on combinatorial and statistical arguments. In Sect. 3.3.4, the most usual criteria are considered. In particular those expressed in [1].

The final result may correspond to a connected—in a single block—or disconnected—in more than one block—seriation. The former connected seriation (resp., the later disconnected seriation) is also called unidimensional (resp., multidimensional) seriation. In the first case (unidimensional), two objects represented by two consecutive rows of the same $\mathcal{O}$ total preorder class or belonging, respectively, to two consecutive classes of the latter preorder possess in common a subset of the Boolean descriptive attributes. Equally, two attributes—represented by two consecutive columns of the data table—in a same $\mathcal{A}$ total preorder class or belonging, respectively, to two adjacent classes of the latter preorder are simultaneously present in a non-empty subset of objects (see Fig. 2.4).

In the case of more than a single block, there is no Boolean attribute simultaneously present into two objects belonging to two distinct blocks. Equally, there is no object possessing two Boolean attributes belonging to two distinct blocks. An idealized figure of a single block is given in Fig. 2.6. Figures 2.5 and 2.8 give image of two blocks. Three blocks are suggested in Fig. 2.10. In each of these cases, the block column set is totally ordered. However, more generally, just a total preorder on the column set of every block is required for the final seriation structure.

The "Block Seriation" method of F. Marcotorchino [2] defines a version of the seriation problem, expressed as a search for a couple $(\mathcal{O}, \mathcal{A})$ of partitions in the same number K of classes which mutually correspond:

$$\pi(\mathcal{O}) = \{O_k | 1 \le k \le K\} \text{ and } \pi(\mathcal{A}) = \{A_k | 1 \le k \le K\} \tag{3.3}$$

In these, the proportion of 1 value in a rectangle of the form $O_k \times A_k$ is "large" while this proportion is "small" in a rectangle of the form $O_h \times A_k$, where h is different from k $(1 \le h \ne k \le K)$.

The criterion defined in this method is *global*. It can be formulated mathematically and has a symmetrical character with respect to the couple $(\mathcal{O}, \mathcal{A})$. It can be perceived visually at the level of the data table. For this reason, we will describe this approach in Sect. 3.2 devoted to the case where the data table is the direct argument of the method concerned.

Another possible approach to "Block Seriation" may be based on Ascendant Hierarchical Clustering. This will be outlined in the same section. Otherwise, let us point out here that different methods in direct joint clusterings of rows and columns of the data table have been considered in the literature [3–6].

To begin, we will attempt to describe in Sect. 3.2.1 the graphical technique of J. Bertin [7] in which the visual perception takes a fundamental part. In this method, the row or column elements of the data table are handled directly.

As just mentioned, the Marcotorchino method—considered in Sect. 3.2.2—may be expressed in terms of combinatorial optimization under constraints. The manipulated argument is an incidence data table composed of zeros and ones. Methodological developments discussing in a relative way, the contribution of the latter method are justified in Sect. 3.2.3.

The Boolean attributes of the incidence data table considered in the Elisséeff heuristic (see Sect. 3.2.4) correspond to categories of binary attributes. This technique is oriented towards ordering the rows by using a global similarity index between each row and the other ones. The order of the columns is adjusted consequently.

In the S. Deutsch and J. Martin technique (see Sect. 3.2.5), each row of the data table is interpreted as a vector of numerical weights. To the latter we will associate the average of the column ranks indicated by Boolean components equal to one. This leads to a similarity notion between rows enabling us to determine an ordering on the rows. A dual interpretation is considered for the columns. In these conditions, an alternating algorithm is proposed.

The evolutionary algorithms constitute a very large optimization algorithm family. In these, the evolution Darwin process is simulated. Starting with a given population, the objective consists of transforming successively the individuals, in order to reach a population composed of efficient individuals. There are three main operations: selection of the best of individuals with respect to a fitness criterion, crossover pairs of individuals and mutations (small modifications) about individuals. All of these three operations are stochastic. The outline of the adaptation of this approach proposed by S. Niermann will be given in Sect. 3.2.6.

In the methods reported in Sect. 3.3, the seriation problem is also posed in terms of combinatorial optimization. Nevertheless, the argument of the algorithm concerned is a similarity matrix S or—equivalently—a dissimilarity matrix $\mathcal{D}$ on the set to be seriated. This set may either be the object set—represented by the data table rows—or the attribute set, represented by the data table columns.

Let E be a finite set—provided with a similarity index S—to be ordered. By denoting m the cardinality of E and by coding E with the sequence $(1, 2, ..., m)$ of the first m integers, the permutation $(\sigma(1), \sigma(2), , \sigma(m))$ to determine on this sequence, in order to satisfy the seriation condition, has to be such that:

$$\sigma(i) < \sigma(j) < \sigma(k) \implies S(i, j) \geq S(i, k) \text{ and } S(j, k) \geq S(i, k) \qquad (3.4)$$

for all 3 subset $\{i, j, k\}$ of $M = \{1, 2, ..., m\}$.

The ordered structure on the set E associated with a seriation might be revealed as an identifiable substructure of a richer structure than a total order on E. For the latter richer structure, we consider in Sect. 3.5 a planar geometrical representation. This has to be as discriminant as possible with respect to the mutual similarities (resp., dissimilarities) between the E elements. In this respect, we will briefly describe in Sect. 3.5.2 the Kendall historical method [8]. This method is intended to organize the object set $\mathcal{O}$ which indexes the row set of the data table. A version of this method starts by building a three-dimensional geometrical representation space according to the *MDSCAL* Kruskal's algorithm [9]. In this representation, one tries to best approximate the system of distance inequalities between the $\mathcal{O}$ elements. The cloud of points so obtained is projected onto the two dimension space sustained by the two first factorial axes of a component principal method (see Sect. 3.5.1). An arrangement of the row points in the order according to a rounded form of a horse-shoe can be observed. It is then necessary to traverse this form, either in one direction or in the opposite one, to recover the order of the points associated with the desired seriation.

The application of the classical factorial analysis methods of principal components and correspondence analyses (see Sect. 3.5.1) will be taken into account in Chap. 4. For these methods, the matter is the representation of the column points corresponding to Boolean attributes, in the geometrical planar space sustained by the two first factorial axes. In the latter, the points are arranged according to a U-shaped or parabolic curve. This type of shape also appears in the geometrical representation provided by the attraction pole method (see Chap. 4, Sect. 4.4.2.2). In Sect. 2.4.6 of Chap. 2, we have shown some affinity between this very simple method and a version of principal component analysis.

For all of the three methods (attraction poles, principal component analysis and correspondence analysis) we may attempt to exploit the dual representation of the $\mathbb{R}^p$ row points on the geometrical plane sustained by the two first factorial axes (eigenvectors). Given the formal constraints of the σ seriation form (see Chap. 2, in Sect. 2.2) we prefer the order on the rows to be derived from the order on the columns in an algorithmic way (see Chap. 5, Sect. 5.4).

3.2 Visual and Combinatorial Methods

3.2.1 Graphical Methods of J. Bertin

For simplicity, let us consider the case for which an incidence data table—composed of zeros and ones—is the argument of the approach concerned. In fact, the latter case is the most frequent for the method. Generally, three stages can be distinguished for the general definition of an algorithm:

1. Choice of the initial step.
2. Progression from the current step to the next one.
3. Stopping rule.

The choice included in the first phase may depend on a certain type of criterion. The development of phase 2 depends also on a criterion of different nature from that considered in 1. Besides, the phase 3 may be followed by a global adequacy criterion different from the two preceding ones.

It is difficult to provide a formal statement for the three phases 1, 2 and 3 in the case of graphical optimal determination of the "matrices ordonnables" as expressed by J. Bertin [7].

It should already be noted that the control of the technique is all the more effective that the incidence data table is of moderate size.

In the manual approach supposed by this technique, the experimenter is facing to his data table. There are three constant features in this method:

1. The progression is carried out from permutations of rows or columns in the data table.
2. The local criteria considered for bringing together rows or columns are informal, they are based on visual perception, they refer to "Visual Data Analysis".
3. A global homogeneity criterion, also based on visual perception, takes part for the evaluation of the final result.

Some remarks Possibly, a theory of the domain might intervene in the local criteria for bringing together rows (resp., columns) of the data table (see 2 above). On the other hand, the quality of the final result get in 3 depends on the clarity of the interpretation of the joint organization of the data table rows and columns.

Now, let us consider the example given in [7] (pages 38 and 39), we shall try to describe the technique by using basic mathematical notions. The procedure is somewhat informal and is in its spirit of recursive nature. First, we assume that there are no missing data. On the other hand, too uniform rows (resp., columns), comprising a large proportion of the same value (0 or 1) are discarded from the analysis. These will be integrated at the end of the process.

The first step of the latter consists of choosing a row which includes a few similar values. For such a row, in statistical terms, the relative frequency of the one (resp., zero) value has to be as near as possible 0.5. Let i_1 designate this row. The latter is placed on the top of the data table.

In these conditions, the different columns whose component $x_{i_1 j}$ (the first one by considering the restructured row order, where j is a generic index for a column) is equal to 1 are taken to the extreme left, according to the following order that we designate by ω_1:

"Column k precedes column l, if and only if the Boolean attribute a^k is more frequent than or equally frequent as a^l, through the row set"

This ordering determines a total preorder on the column set.

The next step consists of bringing to the row i_1—which is on the top—the rows whose profiles are close to the latter row. The proximity perception considered for this purpose is essentially a visual one. It might be relevant here to substitute for the latter proximity intuitive evaluation, a similarity measure between rows of an incidence data table (see Sect. 7.2.2 of Chap. 7 of [6]). This similarity measure may allow a proximity threshold to be numerically specified for the comparison concerned.

Afterwards, we proceed symmetrically with respect to the median row of the data table. To perform this, we carry in the last row, denoted by i_n, that which corresponds to the inverse profile of i_1. Here again we can replace the visual perception evaluation of the difference between i_1 and i_n, by a numerical distance or dissimilarity notion between rows of the incidence data table. In this case, i_n can be defined as the row whose profile is one of those which are at maximum distance from i_1.

In these conditions and symmetrically with respect to above, the different columns, whose last component $x_{i_n j}$ (according to the restructured row order) is equal to 1, are taken to the extreme right, subject to the following order that we designate by ω_n:

"Column k follows column l in the left-right direction column l, if and only if the Boolean attribute a^k is at least as frequent as a^l, through the row set". Otherwise Column k is placed before column l on the right.

ω_1 and ω_n may not be compatible. In this case, a consensus total preorder ω is established in order to rank all the columns of the data table. For this purpose, ω_1 (resp., ω_n) is coded by a ranking function r_1 (resp., r_n) defined on the set $J = \{1, 2, ..., j, ..., p\}$ of the column indices (labels). Then, the ranking function r associated with ω is defined as follows:

$$(\forall j \in J)\ r(j) = \frac{1}{2}\left(r_1(j) + r_n(j)\right) \tag{3.5}$$

The ranking functions r_1 and r_n can be defined adequately with the "mean rank" function (see Sect. 3.3.3 of Chap. 3 of [6]).

Now, by using a row proximity threshold, the rows whose profile is close the last row are selected and grouped downwards.

The process continues in the same way for the central part of the table:

1. Determination of the most distant elements of an ordered pair (l, m) of rows such that l (resp., m) is—in terms of profiles—the nearest the group of rows already stored on the top (resp., at the bottom) of the data table.
2. Associate—by row permutation—the nearest rows of l (resp., m) (we may use here a threshold in relation to a distance index between row profiles).
3. Permute columns which have not been affected previously in order to make more perceptually visible the approximation carried out of the rows.

Repeat recursively 1 to 3 up to a satisfactory overall assessment. The latter may be accompanied by an association coefficient as those developed in Sect. 3.3.

The result obtained is a seriation.

Let us point out that in [10] all the richness of possible graphical manipulations on the data table is detailed. In these, certain blocks are separated in order to highlight a specific structure on the rest.

3.2.2 Block Seriation Method of F. Marchotorchino

Let us return to the problem definition as expressed above (see Sect. 3.1). We consider the description of a set $\mathcal{O}$ of n objects by a set $\mathcal{A}$ of p Boolean attributes (see (2.1) and (2.2)). According to the latter equation, the incidence data table is denoted as follows:

$$T = \{x_i^j | 1 \leq i \leq n, 1 \leq j \leq p\} \tag{3.6}$$

As mentioned above, the aim is to determine a pair $(\pi(\mathcal{O}), \pi(\mathcal{A}))$ of $\mathcal{O}$ and $\mathcal{A}$ partitions, respectively, with the same number K of classes associated, one to one (see (3.3)). The quality of this correspondence between the $\mathcal{O}$ and $\mathcal{A}$ classes is all the better as the proportion of one value in the rectangles of the form $\mathcal{O}_k \times \mathcal{A}_k$, $1 \leq k \leq K$, is large and, on the other hand, this proportion is small in the rectangles of the form $\mathcal{O}_h \times \mathcal{A}_k$, with $h \neq k$ $1 \leq h \neq k \leq K$.

In [2], the problem is posed in terms of maximizing the linear criterion

$$\mathcal{F}(K, \pi(\mathcal{O}), \pi(\mathcal{A})) = \sum_{\{k|1\leq k\leq K\}} \sum_{\{(i,j)\in\mathcal{O}_k\times\mathcal{A}_k\}} x_i^j$$
$$+ \sum_{\{(k,h)|1\leq k\neq h\leq K\}} \sum_{\{(i,j)\in\mathcal{O}_k\times\mathcal{A}_h\}} (1 - x_i^j) \tag{3.7}$$

An heuristic technique based on linear programming in integers (here in zeros and ones) under constraints is employed to maximize (3.7).

It is of importance to notice that a solution to the problem generates $K!$ equivalent solutions which are, respectively, associated with the $K!$ permutations of $(1, 2, ..., k, ..., K)$. More precisely, if $(\tau(1), \tau(2), ..., \tau(k), ..., \tau(K))$ is one of these permutations, the diagonally ordered sequence of blocks

$$(\mathcal{O}_1 \times \mathcal{A}_1, \mathcal{O}_2 \times \mathcal{A}_2, ..., \mathcal{O}_k \times \mathcal{A}_k, ..., \mathcal{O}_K \times \mathcal{A}_K) \tag{3.8}$$

can be permuted according to

$$(\mathcal{O}_{\tau(1)} \times \mathcal{A}_{\tau(1)}, \mathcal{O}_{\tau(2)} \times \mathcal{A}_{\tau(2)}, ..., \mathcal{O}_{\tau(k)} \times \mathcal{A}_{\tau(k)}, ..., \mathcal{O}_{\tau(K)} \times \mathcal{A}_{\tau(K)}) \tag{3.9}$$

By interpreting τ as a bijection on the set $\{1, 2, ..., k, ..., K\}$, the row ordered classes (in the data table) of the partition $\pi(\mathcal{O})$ become

$$\left(\mathcal{O}_{\tau(1)}, \mathcal{O}_{\tau(2)}, ..., \mathcal{O}_{\tau(k)}, ..., \mathcal{O}_{\tau(K)}\right) \tag{3.10}$$

Similarly, the column ordered classes of the partition $\pi(\mathcal{A})$ become

$$\left(\mathcal{A}_{\tau(1)}, \mathcal{A}_{\tau(2)}, ..., \mathcal{A}_{\tau(k)}, ..., \mathcal{A}_{\tau(K)}\right) \tag{3.11}$$

This consideration shows that there is no order notion between the blocks and then the objective is simply to build a pair of partitions on $\mathcal{O}$ and $\mathcal{A}$, respectively, with the same number of associated classes.

3.2.3   Methodological Developments

The main interest of the approach above is to work with respect to a global evaluation criterion defined directly at the level of the entire data table and which can be perceived visually.

In fact, a block seriation decomposition of the incidence data table can be obtained from any clustering method which makes possible to get a significant partition of $\mathcal{A}$ (resp., of $\mathcal{O}$) into a natural number of classes.

To clarify the latter point, suppose we have built the partition $\pi(\mathcal{A})$ considered above (see the second term of (3.3)), by means of a clustering algorithm. Let K denote the number of clusters. A partition $\pi(\mathcal{O})$ into the same number K of classes can be derived by following the Beckner intuitive definition (see Sect. 8.1.1 of Chap. 8 of [6]). More precisely, describing—by means of a loop—the object set $\mathcal{O}$, the object o_i $(1 \leq i \leq n)$ is assigned to the kth cluster $\mathcal{O}_k$ if the relative frequency of the $\mathcal{A}_k$ attributes possessed by o_i is greater than the relative frequency of the $\mathcal{A}_h$ attributes, for h different from k, $1 \leq h \neq k \leq K$. In the case of equality between maximal values of relative frequency presence of different attribute clusters $\mathcal{A}_j$, in the same object o_i, the latter will be assigned to the cluster of objects in constitution whose size is the lowest one and met first.

Dually, from a natural partition $\pi(\mathcal{O})$ into K classes, a partition $\pi(\mathcal{A})$, with the same number of classes can be associated according to the Beckner formulation. More accurate expression of this correspondence is left for the reader.

Now, the question arises to know how to build a natural classification on $\mathcal{A}$ or on $\mathcal{O}$ into a natural number K of classes. Referring to the Likelihood Linkage Analysis (LLA) Ascendant Hierarchical Clustering method, we may retain the partition obtained at the most significant level of the classification tree (see Sect. 9.3.6 of Chap. 9 of [6]).

Otherwise and more generally, different methods associating algorithms with criteria can be envisaged for the aim concerned (see, for example, Chaps. 2, 8, 9 and 10 of [6]).

As emphasized in [11] seriation and clustering are closely related. In fact, these two approaches are fundamentally different. Nevertheless, in data analysis practice, interactions are very fruitful. In these, seriation may follow or precede clustering. In Chap. 7, these interactions will be described. Let us already indicate here some important points which will be taken up in the conclusion Chap. 7.

In the Bertin technique [10], the starting point—prior to the permutation sequence carried out by hand—a pair of permutations on the row and column sets, respectively, are established according to a pair of reduced classification trees. The levels retained for the latter reduction may correspond to those for which a significant node is detected (see Sect. 9.3.6 of Chap. 9 of [6]). Additionally, an interesting technique which may be applied [12–14] consists of permuting the tree leaves of each tree in order to optimize a relational attribute on the set clustered. An example of the latter attribute can be the preordonance associated with the similarity (resp., dissimilarity or distance) index (see Sect. 4.2.2 of [6]) defined on the set to be organized. Here, in our problematic, it is a specific asymmetric structure which is sought from a symmetric proximity matrix, and not from an asymmetric one [15, 16]. In Sect. 6.5 of [17], it is the reverse problem which is considered: induce a hierarchical clustering from a seriation.

Crossing both classification trees (on $\mathcal{A}$ and on $\mathcal{O}$) enables block seriation structure to be revealed. It is of importance to realize that requiring the same number of classes for both decompositions of $\mathcal{A}$ and $\mathcal{O}$ is in fact an immaterial problem. A given cluster in $\mathcal{A}$ (resp., in $\mathcal{O}$) may refer to more than a single cluster in $\mathcal{O}$ (resp., in $\mathcal{A}$).

Nevertheless, let us return to the case for which in the block seriation structure the number of classes is asked to be the same for both decompositions of $\mathcal{A}$ and $\mathcal{O}$, respectively. Whatever the method used to obtain the latter statistical structure, it might be relevant to build, for each of the blocks, a unidimensional ordinal local seriation sustaining the block concerned. More precisely, relative to a given block $\mathcal{O}_k \times \mathcal{A}_k$, we propose to order $\mathcal{A}_k$, $1 \leq k \leq K$. For this, two algorithmic alternatives can be envisaged.

For the first one, we begin by choosing a criterion enabling the detection of an extreme column attribute of $\mathcal{A}_k$ (see Theorems 3 and 4 above). For determining such a column attribute a, belonging to $\mathcal{A}_k$, two associated criteria may be considered, $\mathcal{V}(a)$ and $\mathcal{M}_2(a)$, where $\mathcal{V}(a)$ (resp., $\mathcal{M}_2(a)$) is the variance (resp., the absolute moment of order 2) of the proximities of the $\mathcal{A}_k$ elements to a.

Mostly, the attribute a which maximizes $\mathcal{V}(a)$ is the same as that maximizing $\mathcal{M}_2(a)$. In these cases, the first algorithmic approach is unambiguous. Now, in the case where the attribute a maximizing $\mathcal{V}(a)$ is not identical to that maximizing $\mathcal{M}_2(a)$, two orderings of $\mathcal{A}_k$, associated with the two maximal attributes, can be produced, respectively. The best of them according to a global criterion chosen among those developed in Sect. 3.3 is retained.

The algorithm we refer here to produce an ordering on $\mathcal{A}_k$, obtained from an extreme element of $\mathcal{A}_k$, is that of successive chaining, already expressed in Sect. 3.3. It builds an ordered sequence of the column attributes. The first element of this sequence is the extreme column attribute determined by the maximization of $\mathcal{V}(a)$ or $\mathcal{M}_2(a)$. The kth element of this sequence is that which is the nearest to the extracted

$(k-1)$th element, according to the association coefficient (2.9), $2 \leq k \leq p$. If there are more than one element equally closest to the last extracted element, we may take among these elements the first one.

Let us mention here that more worked out techniques are defined in Sect. 5.3 of Chap. 5. In these, the assignation of the kth element takes into account all of the already seriated sequence of $k-1$ elements.

The second algorithmic approach mentioned above for ordering A_k employs a planar geometrical representation of the set A_k of Boolean attributes. In this representation, the ordered sequence of the A_k elements follows a curve shape. The latter representation can be provided by a factorial analysis method such as principal component analysis or correspondence analysis. More simply, it can be provided by the planar geometrical representation around the two first attraction poles (see Chap. 2, Sect. 2.4.5).

3.2.4 Elisséeff Heuristic

Elisséeff [18–20] developed an algorithmic heuristic of combinatorial and statistical nature apt to bring out a seriation on the object set. This technique is adapted for an incidence data table—crossing an object set with an attribute set—composed of zeros and ones, where a zero value (resp., a one value) indicates a *TRUE* value (resp., a *FALSE* value) of a Boolean attribute on an object.

As usually, the rows of the data table are indexed by objects and the columns, by Boolean attributes. These are defined in the technique proposed from values (categories) of binary attributes. Thereby, the data table corresponds to what is called (in France) "completely disjunctive" data table. In these conditions, let q denote the number of binary attributes. Then $p = 2q$ is the number of the associated Boolean attributes. p is the column number of the incidence data table and—in the case of no missing data—the row vector describing a given object comprises exactly q one values and q zero values. No missing data is assumed in our development.

Let us designate by
$$\mathcal{B} = \{b^j | 1 \leq j \leq q\} \tag{3.12}$$

the set of the binary attributes and by $\{a(j, 1), a(j, 2)\}$ the couple of the two opposite values of b^j, corresponding to two Boolean attributes, $1 \leq j \leq q$. For each of the binary attributes, each of the objects possesses exactly one of its two values. And we have

$$\begin{aligned} &\big((\forall j, 1 \leq j \leq q), (\forall i, 1 \leq i \leq n)\big), \\ &a(j, 1)(i)[1 - a(j, 2)(i)] = 0 \text{ and } a(j, 1)(i) + a(j, 2)(i) = 1 \end{aligned} \tag{3.13}$$

where i indicates the ith object that we designate by o_i, $1 \leq i \leq n$.

Table 3.1 Scalogramme parfait

$\mathcal{O}\backslash\mathcal{A}$	3	5	7	15	13	1	4	6	8	16	14
o_1	X	X	X	X	X	X					
o_2		X	X	X	X	X	X				
o_3			X	X	X	X	X	X			
o_4				X	X	X	X	X	X		
o_5					X	X	X	X	X	X	
o_6						X	X	X	X	X	X

Elisséeff (see Fig. 7 in [18] and Table 3.1) proposed a specific structure of the incidence data table as just described, which he called "scalogramme parfait". The latter is obtained by a row permutation followed by a column one and also by deleting a few number (as small as possible) of objects (individuals) and unusual—in the context—binary attributes.

Now, we shall describe how to bring out the "scalogramme parfait" when the latter underlies the data table. In the procedure considered, there exists a single binary attribute such that one of its two categories (values) is possessed by all of the objects. Let us call c the binary attribute concerned and let $c(1)$ indicate its value, possessed by all of the objects. $c(1)$ will occupy a central position, labeling the median column in the reorganized data table which reveals the "scalogramme parfait".

More precisely, consider the following ordered sequence of the Boolean attributes associated with the categories of the binary attributes:

$$\left(a(1, 1), a(2, 1), ..., a(q-1, 1), \boldsymbol{a(q, 1)}, a(1, 2), a(2, 2), ..., a(q-1, 2), a(q, 2)\right) \tag{3.14}$$

and established such that

$$n(a(1, 1)) < n(a(2, 1)) < ... < n(a(q-1, 1)) < n(\boldsymbol{a(q, 1)}) \tag{3.15}$$

where $n(a(j, 1))$ is the number of objects—and then the number of rows of the incidence data table—for which the Boolean attribute $a(j, 1)$ is *TRUE* $\left(a(j, 1) = 1\right)$.

Necessarily, we have

$$(\forall j, 1 \leq j \leq q), n\left(a(j, 1) + a(j, 2)\right) = n \tag{3.16}$$

where n is the total number of objects, that is, the number of rows in the data table.

In fact, as already taken into account (see (3.13)), we have either $\left(a(j, 1) = 1 \text{ and } a(j, 2) = 0\right)$ or, exclusively, $\left(a(j, 1) = 0 \text{ and } a(j, 2) = 1\right)$.

Considering (3.15), we have

$$n\left(a(1, 2)\right) > n\left(a(2, 2)\right) > ... > n\left(a(q-1, 2)\right). \tag{3.17}$$

Clearly, by considering again (3.15), the Boolean attribute $c(1)$ corresponds to $a(q, 1)$, and thus $a(q, 2)$ is absent from all of the objects, this being expressed by $n\big(a(q, 2)\big) = 0$.

In these conditions, the incidence data table is reduced to $(2q - 1)$ columns, indexed, respectively, by

$$a(1, 1), a(2, 1), ..., a(q - 1, 1), a(q, 1),$$
$$a(1, 2), a(2, 2), ..., a(q - 1, 2) \qquad (3.18)$$

where $a(j, 2)$ is the complementary Boolean attribute of $a(j, 1)$, $1 \leq j \leq q - 1$.

For the "scalogramme parfait" structure of the data table, the row description can be deduced from the definition of the columns where each of them is the vector of the values of a Boolean attribute belonging to $\mathcal{B}$ (see (3.12)).

Now, let us designate the Boolean value set of the data table as follows:

$$\{x_i^j | 1 \leq i \leq q, 1 \leq j \leq 2q - 1\} \qquad (3.19)$$

We have for the ith row of this table:

$$x_i^1 = 0, x_i^2 = 0, ..., x_i^{i-1} = 0$$
$$x_i^i = 1, x_i^{i+1} = 1, ..., x_i^{i+q-1} = 1$$
$$x_i^{i+q} = 0, x_i^{i+q+1} = 0, ..., x_i^{2q-1} = 0 \qquad (3.20)$$

Therefore, there is in all q rows where each of them includes an interval (related sequence) of q entries comprising the one value, the other values of the row concerned being equal to zero. In the first row (resp., in the last row), this interval is situated at the extreme left (resp., the extreme right).

Notice that except the central Boolean attribute $c(1)$, all of the Boolean attributes indexing the columns of the data table correspond—as already expressed—to categories of $(q - 1)$ binary attributes. In the case where the rows and the columns are arranged in order to reveal a "scalogramme parfait", the column vectors from $(q + 1)$ to $(2q - 1)$ can be deduced, one to one, by complementarity from the first $(q - 1)$ column vectors.

According to (3.20), the first binary attribute b^1 is such that $n\big(a(1, 1)\big) = 1$ and $n\big(a(1, 2)\big) = q - 2$. With regard to Table 3.1 for which the label set of the Boolean attributes is

$$\{1, 3, 4, 5, 6, 7, 8, 13, 14, 15, 16\}$$

we may take $a(1, 1) = 3$ and $a(1, 2) = 4$ or $a(1, 1) = 14$ and $a(1, 2) = 13$. In the first case, the column ordering from left to right is

$$(3, 5, 7, 15, 13, 1, 4, 6, 8, 16, 14)$$

and in the second case it is the reverse order, namely,

$$(14, 16, 8, 6, 4, 1, 13, 15, 7, 5, 3)$$

For the latter column order, the row order has to be reversed.

Suppose, as shown in the latter table, which we have chosen $a(1, 1) = 3$. The first column corresponds to the Boolean attribute $a(1, 1)$, possessed by a unique object that we designate by o_1: $a(1, 1)(o_1) = 1$. Then, the row vector description of o_1 is placed in the first row.

The Boolean attribute $a(2, 1)$ is necessarily—according to the "scalogramme parfait" structure—that for which

$$n\big(a(2, 1)\big) = 2 \text{ and } s\big(a(1, 1), a(2, 1)\big) = 1$$

where s designates the raw similarity index. $s\big(a(1, 1), a(2, 1)\big)$ indicates the number of objects for which $a(1, 1)$ and $a(2, 1)$ are *TRUE*.

The second vector column of the Table 3.1 represents the respective values of $a(2, 1)$ and, the second row is associated with the description of the object o_2 different from o_1. For o_2, we have $a(2, 1)(o_2) = 1$.

More generally, assume that for $1 \leq j \leq h$ and h strictly lower than $(q - 1)$ ($h < q - 1$), the columns associated with the different $a(j, 1)$ have already been filled. In these conditions, $a(h + 1, 1)$ will be represented by the $(h + 1)$th column. This will include $(h + 1)$ components equal to one and will be such that

$$s\big(a(h + 1, 1), a(h, 1)\big) = h \qquad (3.21)$$

Therefore, necessarily, the h first components of the $(h + 1)$th column are equal to one and we put the description of the o_{h+1} object in the $(h + 1)$th row. For this object, we have

$$a(h + 1, 1)(o_{h+1}) = 1$$

After putting the first $(q - 1)$ columns, we do equally for the $(q - 1)$ last columns. Recall that the qth column (the central one) is defined by a vector whose components are all equal to one. Therefore, as expressed in (3.18), there are in all $(2q - 1)$ columns. The last $(q - 1)$ columns are defined from the following ordered sequence of Boolean attributes:

$$\big(a(j, 2)|1 \leq j \leq (q - 1)\big) \qquad (3.22)$$

The vector column corresponding to $a(j, 2)$ in (3.22) is complementary to that associated with $a(j, 1)$ in the following sense:

$$\big(\forall j, 1 \leq j \leq (q - 1)\big), a(j, 2)(o_i) = \epsilon \Leftrightarrow a(j, 1)(o_i) = 1 - \epsilon \qquad (3.23)$$

where $\epsilon = 1$ or 0 and where $1 \leq i \leq n$.

Table 3.2 "Scalogramme parfait" and σ form

$\mathcal{A}\backslash\mathcal{O}$	o_1	o_2	o_3	o_4	o_5	o_6
14	x					
16	x	x				
8	x	x	x			
6	x	x	x	x		
4	x	x	x	x	x	
1	x	x	x	x	x	x
13		x	x	x	x	x
15			x	x	x	x
7				x	x	x
5					x	x
3						x

It is in the column $(q + j)$ of the data table that the vector column associated with $a(j, 2)$ is placed.

As announced, we have in fact just described an algorithm which brings out the "scalogramme parfait" structure in the case where the latter underlies the data table. This technique works whatever is the initial column or row ranking.

In this algorithmic description, we have started by ordering the columns, and then we have derived an ordering on the rows, row by row.

In the Elisséeff strategy, we start by ordering the rows and then we deduce the column ordering. This method operates from a real data table for which the form of a "scalogramme parfait" is revealed only after removing a few number of objects and attributes considered archeologically as unusual.

Let us point out here that the object notion as considered above actually covers a set of some individuals having the same description with respect to the retained attributes.

It may be noticed that the "scalogramme parfait" structure corresponds to a strongly chain seriation σ form (see Chap. 2, Sect. 2.4.2), but this occurs in the dual case for which the respective roles of object and attribute are reversed (see Table 3.2).

Now, consider the initial data table expressed as follows:

$$\mathcal{T} = \{\xi_i^j | 1 \leq i \leq n, 1 \leq j \leq 2q\} \tag{3.24}$$

This is a complete disjunctive table of Boolean values, defined by the description of n objects with q binary attributes. We can denote α^{2h-1} and α^{2h} the two opposite Boolean attributes which are defined by the two values of the binary attribute a^h, $1 \leq h \leq q$. Thus, we have

$$(\forall h, 1 \leq h \leq q, \forall i, 1 \leq i \leq n), \xi_i^{2h-1} = \epsilon \Leftrightarrow \xi_i^{2h} = 1 - \epsilon \tag{3.25}$$

where $\epsilon = 0$ or 1.

In the data table considered by Elisséeff (see Fig. 2 of [18]), we have $n = 17$ and $q = 8$. In the technique proposed, the following half square table of raw similarity indices between objects is established:

$$S = \{s(i, i') | 1 \le i < i' \le n\} \qquad (3.26)$$

where

$$(\forall 1 \le i < i' \le n), s(i, i') = \sum_{1 \le j \le 2q} \xi_i^j \cdot \xi_{i'}^j \qquad (3.27)$$

$s(i, i')$ is the number of Boolean attributes which are *TRUE* for i and i'.

In these conditions, to each object is associated the sum of its similarities with the other objects. In more formal words, the following table is built:

$$Ss = \left\{ Ss(i) = \sum \{s(i, i') | 1 \le i' \ne i \le n\} \right\} \qquad (3.28)$$

$Ss(i)$ is called "richness" of i, $1 \le i \le n$. In these conditions, the richest element is detected. The latter will play a fundamental role in ordering the rows of the data table, thus to bring out the "scalogramme parfait" structure. In fact, the richest row is coupled with a row element as rich as possible. The case where several elements are equally the richest is not developed. However, an adaptation of the procedure must be possible.

Let r designate the row index of the richest element and let t be the index of a richest element after r (t is the richest element of all rows except r). Consider now the ordered sequence of the rows, established with respect to r in the following form:

$$\big(r(1), r(2), ..., r(i), ..., r(n-2)\big) \qquad (3.29)$$

such that

$$s(r, r(1)) \ge s(r, r(2)) \ge ... \ge s(r, r(i)) \ge ... \ge s(r, r(n-2)) \qquad (3.30)$$

We start by placing the row r above the row t. Generally, at a given step, following the order defined in (3.29), each of the remaining rows (not yet placed) is compared, on the one hand, with the last element placed above r and, on the other hand, with the last element placed below t. The strongest association according to the similarity index **s** determines the position of the element compared: directly with the last element placed above r or directly under the last element placed below t.

In the first step, each of the elements is compared to r and to t, in the order according to (3.29). The first element—generally $r(1)$—for which inequal similarities are observed has to be above r or below t. In the case of equal similarities, the assignation of the element concerned remains waiting for a pair of discriminated elements in their respective associations with r and t.

In these conditions, the pending element is compared with each of the two components of the discriminated pair. It is then placed on the side of the component of which it is the closest. In this placement, its similarity value with r (if it is to be placed above r) or with t (if it is to be placed below t) is taken into account. For more details, refer to the example given in [18] page 7, Fig. 4.

In [18], the whole approach is described by using a real case, where the matter consists of a sample of Chinese archaïc bronzes of type Yeou large assigned to a period between 1400 and 800 B.C. We have given above more mathematical formulation which reflects more clearly the nature of the procedure. Nevertheless, the basic algorithm that determines the row ranking must be much more formalized and based mathematically.

The Elisséeff heuristic is essentially combinatorial and statistical in nature. In this respect, we are very surprised to observe that in [11] Bertin's approach is considered as having a descent link with that of Elisséeff. While, in fact, the Bertin procedure is essentially based on visual perception.

The new methods proposed in this book are combinatorial and statistical in nature. They were conceived completely independent of the Elisséeff approach. In fact, it is only on the occasion of writing this book that we have learned the latter approach.

The origin of our approach is based on the analysis of the variance of statistical proximities. We designed it to measure the degree of neutrality of each element of the set to be organized in relation to a clustering objective (see [3, 6, 21])

Very generally—as regards the seed idea—for the approach of Elisséeff, one seeks to determine a median row vector from which it is necessary to arrange, on both sides the other rows, while for us—in the simplest case —we determine an extreme column, in relation to which we rank on the same side the other columns. Our method is less ambiguous and more stable.

Our contribution to the seriation subject has already given rise to [22–24].

3.2.5 The Deutsch and Martin Algorithm

It is only recently we learned about the algorithm of S. Deutsch and J. Martin [25]. Starting from an arbitrary initial arrangement, this algorithm has an alternating character up to desired convergence, passing from a permutation of the elements of one side of the array, rows or columns, to the other side of the array, columns or rows, respectively.

Suppose, as it was often the case in this book, that the data table is an incidence data table composed of zeros and ones, associated with Boolean data. Note that in [25], extension is expected from this data structure to any array of positive real numbers resulting from description of a set of objects by numerical variables. In the latter case, a scaling is needed. This is not clarified in the reference mentioned.

In this algorithm, the row set and the column set play symmetrical roles, each with respect to the other. Now, let us refer to the data table $\mathcal{T}$ (see (2.2)) that we take up again here

$$T = \{x_i^j | 1 \le i \le n, 1 \le j \le p\} \tag{3.31}$$

Suppose that the first step consists of ordering the row set. For this purpose, each of the rows is interpreted as an empirical probability distribution on the set of the column ranks $\{1, 2, ..., j, ..., p\}$. More precisely, the probability distribution defined by the ith row that we designate by f_i is written as

$$f_i = \left(\frac{x_i^j}{\sum_{1 \le h \le p} x_i^h} | 1 \le j \le p \right) \tag{3.32}$$

Therefore, we will associate with each row i the average $\overline{x_i}$ of the column ranks as follows:

$$\overline{x_i} = \frac{\sum\{j.x_i^j | 1 \le j \le p\}}{\sum\{x_i^h | 1 \le h \le p\}} \tag{3.33}$$

$1 \le i \le n$.

In this way, the proximity between two values $\overline{x_i}$ and $\overline{x_{i'}}$ induces, in a certain way, similarity between the ith and the i'th rows. In these conditions, we order the rows according to increasing values of $\overline{x_i}$ (see (3.33)), $1 \le i \le n$. More explicitly, if $(x_{i(1)}, x_{i(2)}, ..., x_{i(n)})$ is the row permutation concerned, we have

$$\overline{x}_{i(1)} \le \overline{x}_{i(2)} \le ... \le \overline{x}_{i(n)} \tag{3.34}$$

Then, we proceed in the same way to order the columns in relation to the rows of the new table. By denoting it in the same way as previously, the criterion takes the form

$$\overline{x^j} = \frac{\sum\{i.x_i^j | 1 \le i \le n\}}{\sum\{x_l^j | 1 \le l \le n\}} \tag{3.35}$$

The two steps are then repeated until to reach a stationary state.

An initial arrangement of the array can be obtained from a pair of permutations randomly defined on the rows and columns, respectively. When, in real cases, we consider different initializations, we do not necessarily get the same solution to convergence, but solutions very close or very similar, some of them being identical. In this case, for each of the rows, corresponding to a given object, the average of the rank positions it has occupied in the different solutions is calculated (the range of possible positions goes from 1 to n). Similarly, for each of the columns representing a given attribute, we calculate the average of the rank positions it has occupied in the different solutions (the range of the respective positions goes from 1 to p). In these conditions, the rows and the columns are ordered, respectively, according to the values of these two averages.

One final point. In the case of large data table, the algorithm can cycle around a small number of solutions. We can adopt one of them or change the initialization.

Such phenomena (several stable solutions obtained from different initializations, cycling around some solutions for a given initialization) do not really occur for the other methods whether geometric or combinatorial methods.

Recall that in our approach, rows and columns of the data table are considered asymmetrically. A given column is defined by the sequence of the respective values of different descriptive attributes on a same object. We start by building an ordering on the attribute set (represented by the columns) and then we derive an associated order (see Chap. 5, Sect. 5.2). Notice that the latter order is obtained in the case of an incidence data table by using an index as in (3.33) (see Chap. 5, Sect. 5.2.3). This technique was proposed and performed independently of the contribution of [25] that we did not know when we propose our technique. And, in fact, our method is essentially different; it is not an alternated. On the other hand, our formal presentation provides a rationale to the Deutsch and Martin way of doing.

3.2.6   The Niermann Algorithm

The solution proposed by S. Niermann [26] for the seriation problem consists of an adaptation of the genetic algorithmic also said evolutionary computation. In the case of an incidence data table composed of zeros and ones, the objective consists of determining an ordered pair of permutations, on the rows and columns of this table, respectively, in order to reveal a diagonal form for which the density of one value is as high as possible.

As mentioned in the introduction of this chapter (see Sect. 3.1), the evolutionary algorithms constitute a very large family of optimization algorithms. Starting with a given population, the process consists of transforming successively the individuals of this population in order to reach a population composed of efficient individuals. There are three main operations to define for this transformation: *crossover* pairs of individuals, *mutation* to apply to individuals and *selection* of the best individuals in relation to a fitness criterion. All of these three operations have a stochastic character.

Initially, genetic algorithms were developed in the framework of Boolean descriptions. In these, the unit element is a Boolean vector [27, 28]. Since then, they have expanded enormously. In [29], a very important set of applications in economics is presented. Our own experience in favor of R. Ngouënet PhD [30] deals with the geometric representation of a dissimilarity defined on a finite set. This concerns the problem of multidimensional scaling [31]. It is a parallel implementation of the algorithm which has been carried out [32].

Now, let us give the description of the adaptation of evolutionary (genetic) algorithms to the seriation problem as given by S. Niermann [26]. In this adaptation, a given individual of the start population is defined by the initial data table, but to which two random permutations have been applied: one on the rows and the other on the columns. In these conditions, statistically, the total population or, in other words, the search space includes $n! \times p!$ elements. From this, a subset of m distinct elements, randomly drawn, defines the initial population in the meaning of evolutionary computation (here, an element is an array where rows and columns have been permuted independently). In the example given in the article, $n = 9$, $p = 16$ and $m = 20$. The latter population—comprising m elements—will evolve from three operations applied iteratively: *crossover*, *mutation* and *selection*.

Notice that the size of the random subset defining the initial genetic population ($m = 20$) is very small compared to the search space cardinality ($9! \times 16!$). A more reasonable solution to the optimization problem would consist of

1. Randomly drawn, without replacement, a large sample defining a subset E of the search space.
2. Assign to E a fitness function as in (3.50).
3. Randomly drawn with replacement, according to the fitness function, the number of times needed to obtain 20 elements.

Anyway, we will refer in what follows to a sample sized 20 considered as the initial population. The latter will evolve from crossover, mutation and selection operations, each of them having a stochastic character.

The set of the individuals can be denoted as follows:

$$\mathcal{I} = \{(\pi(\underline{r}_k), \pi(\underline{c}_k)) | 1 \leq k \leq 20\} \tag{3.36}$$

where $\underline{r}_k$ (resp., $\underline{c}_k$) represents a permutation of the rows (resp., of the columns) of the array. We can designate these two permutations as follows:

$$\pi(\underline{r}_k) = (i_1, i_2, ..., i_n)$$
$$\text{and}$$
$$\pi(\underline{c}_k) = (j_1, j_2, ..., j_p) \tag{3.37}$$

Unordered pairs of the array can be written as follows:

$$\mathcal{P} = \left\{ \{(\pi(\underline{r}_k), \pi(\underline{c}_k)), (\pi(\underline{r}_l), \pi(\underline{c}_l))\} | 1 \leq k < l \leq 20 \right\} \tag{3.38}$$

Crossover operation

As stated above, we refer to a population comprising 20 individuals. In the latter, 20 pairs of distinct individuals are randomly chosen. The probability that a given pair be subject to a crossover is fixed at 0.5. And then, the number of pairs subject to a crossover is a random binomial variable parametrized by $(20, 0.5)$. If

$$\{(\pi(\underline{r}_k), \pi(\underline{c}_k)), (\pi(\underline{r}_l), \pi(\underline{c}_l))\} \tag{3.39}$$

is a pair giving a crossover, its two offsprings are

$$\{(\pi(\underline{r}_k), \pi(\underline{c}_l)), (\pi(\underline{r}_l), \pi(\underline{c}_k))\} \tag{3.40}$$

Mutation operation

Consider an individual $\left(\pi(\underline{r}), \pi(\underline{c})\right)$ subject to a mutation. This mutation transforms both $\pi(\underline{r})$ and $\pi(\underline{c})$. For each individual in the population which includes 20 elements, the author [26] sets the probability to be subject to a mutation at 0.5. In this case, two random and independent segments, belonging, respectively, to $\pi(\underline{r})$ and $\pi(\underline{c})$ are inverted. More precisely, denote $\pi(\underline{r})$ and $\pi(\underline{c})$ as follows:

$$\pi(\underline{r}) = (i_1, i_2, ..., i_n)$$
$$\text{and}$$
$$\pi(\underline{c}) = (j_1, j_2, ..., j_p) \tag{3.41}$$

Denote also

$$(i_a, i_{a+1}, ..., i_b) , 1 \leq a < b \leq n$$
$$\text{and}$$
$$(j_c, j_{c+1}, ..., j_d) , 1 \leq c < d \leq p \tag{3.42}$$

the two independent random segments obtained, respectively, in $\pi(\underline{r})$ and $\pi(\underline{c})$. The mutation will transform these into

$$\pi(\underline{r_{mut}}) = (i_1, i_2, ..., i_{a-1}, i_b, i_{b-1}, ..., i_a, i_{a+1}, ..., i_n)$$
$$\text{and}$$
$$\pi(\underline{c_{mut}}) = (j_1, j_2, ..., j_{c-1}, j_d, j_{d-1}, ..., j_c, j_{c+1}, ..., j_p) \tag{3.43}$$

respectively.

Let us notice that the random determination of a segment assumes the random determinations of its two limits, independently.

Selection operation

At a given stage, a population comprising m elements is available ($m = 20$ in the article concerned). The elements are not necessarily distinct at a current stage. However, they are mutually distinct at the first stage. As mentioned above, nothing is said about the initial population of m elements in [26].

Selection results from a random drawn with replacement from the population of the current stage. m individuals are extracted according to their respective fitnesses, numerically measured as follows.

We will associate with a given data table its *STRESS*. This is interpreted as a measure of disorder in relation to a seriation structure of the data array.

$$STRESS = \sum_{i=1}^{n} \sum_{j=1}^{p} d^2(i,j) \qquad (3.44)$$

where

$$d^2(i,j) = \sum_{r=max(1,i-1)}^{min(n,i+1)} \sum_{c=max(1,j-1)}^{min(p,j+1)} (x_{ij} - x_{rc})^2 \qquad (3.45)$$

In these conditions, the fitness associated with a data table $\mathcal{T}$ takes the following form:

$$fitness(\mathcal{T}) = \sum_{i=1}^{n} \sum_{j=1}^{p} a(i,j) \qquad (3.46)$$

where

$$a(i,j) = [min(n, i+1) - max(1, i-1)] \times [min(p, j+1) - max(1, j-1)] - d(i,j) \qquad (3.47)$$

Therefore, we consider on the following set of individuals:

$$\{\mathcal{T}_k | 1 \le k \le m\} \qquad (3.48)$$

The following numerical distribution of numerical positive values of sum 1:

$$\{\varphi_k | 1 \le k \le m\} \qquad (3.49)$$

where

$$\varphi_k = \frac{fitness(\mathcal{T}_k)}{\sum_{1 \le h \le m} fitness(\mathcal{T}_h)} \qquad (3.50)$$

According to (3.49), interpreted as a probability distribution, the random drawn with replacement can be performed in order to produce a new population. Some of the individuals of preceding population can appear more than once. The three steps (crossover, mutation and selection) are carried out iteratively until a predetermined number of iterations is reached or until the convergence of the maximum value of the fitness in the population.

Classically, before a next iteration, the individual with the highest fitness in the current population is retained if it is better than the last retained one.

In [26], the selection operation wishes to be based on tournaments. The set of unordered pairs of individuals in the population is considered. For a given pair, the component defined by the individual with the highest fitness is likely to be retained for the next generation. For a population of 20 individuals, there are 190 unordered pairs of individuals. In the latter set of these, 20 pairs are chosen with replacement and each of these pairs perform a tournament. Notice that 20 is small compared to

190. On the other hand, we cannot know the number of distinct individuals which will appear in the random chosen pairs of individuals. Moreover, does the fitness notion—as defined in (3.50)—include all the possible arrays? A mathematical formulation could have clarified this type of technique.

3.3 Seriation Defined as a Combinatorial Optimization Problem

3.3.1 Preamble

The title of this section is exactly that of Sect. 2 of [33]. In the latter—as the case is for most presentations, the seriation problem is addressed to order directly the object set $\mathcal{O}$ and not the attribute set $\mathcal{A}$. Moreover, this problem is directly posed in terms of combinatorial methods concerning data defined by a symmetrical dissimilarity matrix $\mathcal{D}$. This case is known as that of two-way one-mode data. These methods can easily be transposed for ordering the attribute set $\mathcal{A}$, in the situation where $\mathcal{A}$ is endowed with an association coefficient such as (2.9). This adaptation will be indicated in the following.

In fact, four cases can be envisaged depending on whether the set to be seriated is the object set $\mathcal{O}$ or the attribute set $\mathcal{A}$, and, on the other hand, according to whether the set to be seriated is provided with an index of dissimilarity or similarity. Thus, the symmetrical square matrix of the data is indexed either by $\mathcal{O} \times \mathcal{O}$ or by $\mathcal{A} \times \mathcal{A}$. In the case where the description table $\mathcal{O} \times \mathcal{A}$ of the data is provided, it is necessary to establish one or the other of these latter matrices.

As already claimed, at the end of the introductory section and also, just above, our approach consists first of ordering the set $\mathcal{A}$ of column attributes. Ordering the set $\mathcal{O}$ of row objects is then deduced (see Chap. 5, Sect. 5.4).

3.3.2 General Facets

To begin, we shall consider the seriation of an object set $\mathcal{O}$ provided with a dissimilarity matrix $\mathcal{D}$. Now, let us specify $\mathcal{D}$ as follows:

$$\mathcal{D} = \{D(i, j) | 1 \leq i, j \leq n\} \tag{3.51}$$

The problem is posed in terms of seeking for a permutation ρ of the sequence $(1, 2, ..., i, ..., n)$ such that the numerical binary valuation of

$$\mathbb{I}^{[2]} = \{(i, j) | 1 \leq i \neq j \leq n\} \tag{3.52}$$

defined by

$$\rho(\mathcal{D}) = \{D(\rho(i), \rho(j)) | (i, j) \in \mathbb{I}^{[2]}\} \tag{3.53}$$

is as close as possible—in terms of association—with the numerical binary valuation

$$\mathbb{E} = \{e(i, j) | (i, j) \in \mathbb{I}^{[2]}\} \tag{3.54}$$

where $e(i, j)$ is an increasing function of $|i - j|$, $1 \le i \ne j \le n$. Commonly, the following two functions are considered:

$$(\forall (i, j) \in \mathbb{I}^{[2]}), e(i, j) = |i - j|$$

and

$$(\forall (i, j) \in \mathbb{I}^{[2]}), e(i, j) = |i - j|^2$$

For this purpose, we may either maximize a similarity function or minimize a dissimilarity one between the binary valuations (3.53) and (3.54). In our development below, only the dissimilarity version of the association will be considered. The similarity one can be easily derived (see (3.59) and (3.60)). A raw association coefficient between $\rho(\mathcal{D})$ and $\mathbb{E}$ can be written as follows:

$$\mathbf{s}(\rho(\mathcal{D}), \mathbb{E}) = \sum_{(i,j) \in \mathbb{I}^{[2]}} D(\rho(i), \rho(j)).e(i, j) \tag{3.55}$$

In terms of an inertia criterion [34], an association function is proposed in the following form:

$$\mathbf{t}(\rho(\mathcal{D}), \mathbb{E}) = \sum_{(i,j) \in \mathbb{I}^{[2]}} D(\rho(i), \rho(j)).(|i - j|^2) \tag{3.56}$$

In our method (see Sect. 3.3.3), we will mainly be interested in the case for which $e(i, j) = |i - j|$, $1 \le i \ne j \le n$. In these conditions, the association coefficient (3.55) becomes

$$\mathbf{s}(\rho(\mathcal{D}), \mathbb{E}) = \sum_{(i,j) \in \mathbb{I}^{[2]}} D(\rho(i), \rho(j)).|i - j| \tag{3.57}$$

It is of importance to notice that a variable change in the summation shows that the association coefficient (3.55) can be written as follows:

$$\mathbf{s}(\rho(\mathcal{D}), \mathbb{E}) = \sum_{(i,j) \in \mathbb{I}^{[2]}} D(i, j).e(\tau(i), \tau(j)) \tag{3.58}$$

where τ is the inverse permutation of ρ.

Let us illustrate here this point. For $\mathbb{I} = \{1, 2, 3, 4, 5, 6, 7, 8\}$ and

$$\rho(\mathbb{I}) = \big(\rho(1), \rho(2), \rho(3), \rho(4), \rho(5), \rho(6), \rho(7), \rho(8)\big) = \big(5, 1, 7, 2, 4, 6, 8, 3\big),$$

we have

$$\tau(\mathbb{I}) = \big(\tau(1), \tau(2), \tau(3), \tau(4), \tau(5), \tau(6), \tau(7), \tau(8)\big) = \big(2, 4, 8, 5, 1, 6, 3, 7\big)$$

Concerning the matrix expression of $\rho(\mathcal{D})$ (see (3.53)), it is of importance to notice that the ith row (resp., jth column) of the latter is the $\rho(i)$th row (resp., $\rho(j)$th column) of the initial matrix $\mathcal{D}$ (see (3.51)). Thereby, $D(\rho(i), \rho(j))$ has to be associated with $e(i, j)$, $1 \leq i, j \leq n$.

Now, in relation to (3.58) for the reordered matrix $\tau(\mathbb{E})$, the row $\tau(i)$ (resp., the column $\tau(i)$) includes the former i row (resp., i column) of the initial matrix $\mathcal{D}$ (see (3.51)). In these conditions, $D(i, j)$ is the distance value met at the intersection of the ith row and the jth of the reordered matrix indexed by $\tau(\mathbb{I}) \times \tau(\mathbb{I})$. Therefore, $D(i, j)$ has to be associated with $e(\tau(i), \tau(j))$, $1 \leq i, j \leq n$.

Now, if we assume that the given data is a similarity matrix $\mathcal{S}$

$$\mathcal{S} = \{S(i, j) | (i, j) \in \mathbb{I}^{[2]}\} \tag{3.59}$$

the association criterion proposed in [35] takes the following form:

$$\mathbf{u}(\mathcal{S}, \ll(\mathbb{I})) = \sum_{(i,j) \in \mathbb{I}^{[2]}} S(i, j).(\tau(i) - \tau(j))^2 \tag{3.60}$$

In this case, the matter consists of seeking for a permutation τ which minimizes $\mathbf{u}(\mathcal{S}, \ll(\mathcal{E}))$.

As just expressed, the fundamental nature of this problem is optimization of an association coefficient between the valued binary relations (3.53) and (3.54) with respect to the parameter permutation ρ. In fact, this assignment problem is very general. An *LLA* association coefficient is developed in Sect. 6.2.5 of Chap. 6 of [6] for comparing (3.53) and (3.54). There are several specific optimization problems for comparing two structured relations (see page 184 of [36]). A version in which the variable parameter is a permutation has been considered in pattern recognition (see Chap. 10 of [21]).

The methods considered here—in connection with Sect. 2 of [35]—refer to a *local* optimization of a *global* criterion. We have indicated above by $\mathbf{G}$ (see Sect. 3.1) this type of methods. We have also expressed that for our own, we do prefer an approach of type $\mathbf{L}$ corresponding to optimizing a *local* criterion. Nonetheless, as mentioned in Sect. 3.1, a global criterion can be employed in order to assess the result obtained by a local one. This will be carried out in Chap. 5 by using the *LLA* association coefficient (see Sect. 3.3.3).

Clearly, as just mentioned, a combinatorial optimization approach established to associate (3.53) and (3.54) can be transposed in the context of ordering $\mathcal{A}$. This is the case that will interest us first (see Sect. 3.3.3). The data considered is an incidence data table crossing an object set $\mathcal{O}$ with an attribute set $\mathcal{A}$. $\mathbb{J} = \{1, 2, ..., k, ..., l, ..., p\}$ indexing the column attribute set, we have to establish the symmetrical square distance matrix

$$\mathcal{D}(\mathbb{J}) = \{D(k, l)|1 \leq k, l \leq p\} \qquad (3.61)$$

where $D(k, l)$ can be derived from (2.9) or, equivalently, from (2.12). More precisely, we have

$$D(k, l)^2 = S(k, k) + S(l, l) - 2S(k, l) = \sum_{1 \leq i \leq n} (x_i'^k - x_i'^l)^2$$

More elaborate dissimilarity table directly related to the *Likelihood Linkage* approach is given by

$$\mathcal{D}_{inf}(\mathbb{J}) = \{D_{inf}(k, l)|1 \leq k \neq l \leq p\} \qquad (3.62)$$

where this matrix is that of *informational* dissimilarities expressed in (10.1.6) of Chap. 10 of [6]. However, by considering the nature of the problem posed, the dissimilarity matrix $\mathcal{D}$ given in (3.61) and specified just above will be sufficient for our analysis.

Therefore, in order to evaluate a permutation $\rho(\mathbb{J})$ on $\mathbb{J}$, we have to match

$$\{D(\rho(k), \rho(l))|(k, l) \in \mathbb{J}^{[2]}\}$$
$$\text{with}$$
$$\{e(k, l)|(k, l) \in \mathbb{J}^{[2]}\} \qquad (3.63)$$

where $\mathbb{J}^{[2]} = \{(k, l)|1 \leq k \neq l \leq p\}$ and where $e(k, l)$ is an increasing function of $|k - l|$.

By its very nature, the analysis does not need to consider a developed function e and then, as mentioned above, we will consider the easiest one in our analysis, that is to say:

$$e(k, l) = |k - l| , (k, l) \in \mathbb{J}^{[2]}$$

At the risk of repetition we shall use the same type of illustration as above, but this time, relative to $\mathbb{J}$ which indexes the set $\mathcal{A}$ of the column attributes. This option is somewhat related to the fact that the first step of our method consists of ordering $\mathbb{J}$. Here, we will give a more substantial development of the correspondence between the initial distance matrix (3.61) and that reordered according to the seriation hypothesis (see (3.63)).

In this illustration, $\mathbb{J}$ comprises five elements and it can be denoted as follows:

$$\mathbb{J} = \{1, 2, 3, 4, 5\}$$

Now, let us consider the following permutation:

$$\rho(1, 2, 3, 4, 5) = \big(\rho(1), \rho(2), \rho(3), \rho(4), \rho(5)\big) = (3, 5, 4, 1, 2)$$

ρ can be interpreted as a self-bijection of $\mathbb{J}$, substituting the sequence $(3, 5, 4, 1, 2)$ for $(1, 2, 3, 4, 5)$.

As above, let us designate by τ the ρ inverse permutation ($\tau = \rho^{-1}$). We have

$$\tau(1, 2, 3, 4, 5) = \big(\tau(1), \tau(2), \tau(3), \tau(4), \tau(5)\big) = (4, 5, 1, 3, 2)$$

We shall focus on the image by ρ, first of the dissimilarity matrix $\mathcal{D}(\mathbb{J})$ considered in (3.61) and next of the following similarity matrix on $\mathbb{J}$:

$$S(\mathbb{J}) = \{S(k, l) | 1 \le k, l \le p\} \tag{3.64}$$

According to the above notation (see (3.63)), let us designate by $\mathcal{D}(\rho(\mathbb{J}))$ and $S(\rho(\mathbb{J}))$ these images. We have

$$\mathcal{D}(\rho(\mathbb{J})) = \{D(\rho(k), \rho(l)) | 1 \le k, l \le p\} \tag{3.65}$$

$$S(\rho(\mathbb{J})) = \{S(\rho(k), \rho(l)) | 1 \le k, l \le p\} \tag{3.66}$$

In these matrices, the row (resp., the column) numbered k is occupied by the row (resp., the column) $\rho(k)$ of the initial matrix. Otherwise, by reasoning with respect to the permutation τ, the row (resp., the column) $\tau(k)$ is occupied by the former row k (resp., column k).

In the matrix $\mathcal{D}(\rho(\mathbb{J}))$, the dissimilarity value $D(\rho(k), \rho(l))$ occurs at the intersection of the row occupied by $\rho(k)$ and the column occupied by $\rho(l)$. This value is—in our option—to match with $e(k, l) = |k - l|$. More generally, it has to be matched with $\varphi(e(k, l))$, where φ is an increasing function.

Let us reason now in terms of the permutation τ. For this, consider the ordered pair (k, l) of $\mathbb{J} \times \mathbb{J}$ such that the initial row k (resp., column k) occupies the row $\tau(k)$ (resp., column $\tau(k)$). And, the initial row l (resp., column l) occupies the row $\tau(l)$ (resp., column $\tau(l)$). We suppose that the ordered sequence of the rows, indexed by $(1, 2, ..., j, ..., p)$. Thus, the jth row (resp., column) contains the row h (resp., column h) of the initial matrix for which $\tau(h) = j$. Thereby, at the intersection of the row $\tau(k)$ and the column $\tau(l)$, we have the distance value $D(k, l)$. The latter has to be put in correspondence with $\varphi(e(\tau(k), \tau(l)))$, where φ is an increasing function. By considering the example above for $\tau(3) = 1$ and $\tau(1) = 4$, $D(1, 3)$ occurs at the intersection of the row $\tau(3)$ and the column $\tau(1)$. $D(1, 3)$ has to be matched with $e(1, 4)$.

The distance matrix of Table 3.4 results from Table 3.3 after applying the ρ permutation on its row set (resp., the column set).

By considering the *upper* half distance matrix including the diagonal, of Table 3.4, we observe that the sequence of the distance values from left to right on a same row,

Table 3.3 Dissimilarity matrix—1

$\mathbb{J}\backslash\mathbb{J}$	1	2	3	4	5
1	0.	0.5	0.9	0.5	0.85
2	0.5	0.	1.	0.85	0.9
3	0.9	1.	0.	0.6	0.4
4	0.5	0.85	0.6	0.	0.5
5	0.85	0.9	0.4	0.5	0.

Table 3.4 Dissimilarity matrix—2

$\mathbb{J}\backslash\mathbb{J}$	3	5	4	1	2
$\rho(1) = 3$	0.	0.4	0.6	0.9	1.
$\rho(2) = 5$	0.4	0.	0.5	0.85	0.9
$\rho(3) = 4$	0.6	0.5	0.	0.5	0.85
$\rho(4) = 1$	0.9	0.85	0.5	0.	0.5
$\rho(5) = 2$	1.	0.9	0.85	0.5	0.

from the diagonal, increases. This property can expressed by

$$\big(\forall(k, l), 1 \leq k \leq l < p\big), D(\rho(k), \rho(l)) \leq D(\rho(k), \rho(l + 1)) \tag{3.67}$$

Equally, we observe the sequence of the distance values from top to bottom, till the diagonal, decreases. This property can be expressed by

$$\big(\forall(k, l), 1 < k \leq l \leq p\big), D(\rho(k), \rho(l)) \geq D(\rho(k + 1), \rho(l)) \tag{3.68}$$

Note that in these latter two formulas $\rho(k)$ is the row index and $\rho(l)$ the column one.

In the case where both conditions (3.67) and (3.68) are satisfied, the permuted distance matrix is said to have an anti-Robinson structure [37]. The case of a Robinson structure concerns a similarity matrix as that of Table 3.6, which results from Table 3.5. In this case, relative to the *lower* semi-matrix, we have

$$\big(\forall(k, l), 1 < k \leq l \leq p\big), S(\rho(k), \rho(l)) \geq S(\rho(k - 1), \rho(l)) \tag{3.69}$$

and

$$\big(\forall(k, l), 1 \leq k \leq l < p\big), S(\rho(k), \rho(l)) \geq S(\rho(k), \rho(l + 1)) \tag{3.70}$$

Note that in these latter two formulas $\rho(k)$ is the column index and $\rho(l)$ the row one.

Thereby, in the case where there exists a permutation ρ for which the dissimilarity matrix $\mathcal{D}(\rho(\mathbb{J}))$ (resp., the similarity matrix $\mathcal{S}(\rho(\mathbb{J}))$) satisfies the conditions (3.68)

Table 3.5 Similarity matrix—1

$\mathbb{J}\backslash\mathbb{J}$	1	2	3	4	5
1	1.	0.5	0.1	0.5	0.15
2	0.5	1.	0.	0.15	0.1
3	0.1	0.	1.	0.4	0.6
4	0.5	0.15	0.4	1.	0.5
5	0.15	0.1	0.6	0.5	1.

and (3.69) (resp., the conditions (3.70) and (3.70)), then ρ determines a seriation on $\mathbb{J}$ without any deviation.

In real cases when a given seriation is strongly justified, statistical reasons make that differences can be observed with respect to the ordering sustaining the seriation concerned. In these conditions, the matter consists of determining a permutation ρ which enables the best fitting to be realized with respect to a seriation. Once again, given our method, we consider the latter on $\mathbb{J}$. Moreover, $\mathbb{J}$ is assumed provided with a dissimilarity matrix $\mathcal{D}(\mathbb{J})$ (see (3.61)). In the basic case considered above where the data is an incidence matrix crossing objects with attributes, the distance $D(k, l)$ is calculated according to the equation which follows directly (3.61), $1 \leq k, l \leq p$.

Taking into account the preceding considerations, the criterion to be maximized for a better quality of the permutation ρ giving rise to a seriation is (see (3.57) and (3.56)) that we take up again here for the ordering of $\mathbb{J}$

$$\mathbf{s}(\rho(\mathcal{D}), \mathbb{E}) = \sum_{(k,l)\in\mathbb{J}^{[2]}} D(\rho(k), \rho(l)).|k - l| \tag{3.71}$$

We could also choose as criterion (see (3.56)):

$$\mathbf{t}(\rho(\mathcal{D}), \mathbb{E}) = \sum_{(k,l)\in\mathbb{J}^{[2]}} D(\rho(k), \rho(l)).(k - l)^2 \tag{3.72}$$

Each of the latter both measures can be viewed as a raw association coefficient between two binary numerical valuations on $\mathbb{J}$. In fact, to realize the quality of a seriation associated with a permutation ρ, we will consider, instead of the raw index (see (3.71) and (3.72)), a statistically normalized coefficient consistent with the optics of the *LLA* approach. This has already been mentioned above when the valued relations are defined on $\mathbb{I}$. The detail of the components of this coefficient will be expressed in Sect. 3.3.3.1.

We leave for the reader to provide an analogous development to that made in Sect. 5.4, but in the case where the data with respect to which we work is a similarity matrix $\mathcal{S}(\mathbb{J})$.

Table 3.6 Similarity matrix—2

$\mathbb{J}\backslash\mathbb{J}$	3	5	4	1	2
$\rho(1) = 3$	1.	0.6	0.4	0.1	0.
$\rho(2) = 5$	0.6	1.	0.5	0.15	0.1
$\rho(3) = 4$	0.4	0.5	1.	0.5	0.15
$\rho(4) = 1$	0.1	0.15	0.5	1.	0.5
$\rho(5) = 2$	0.	0.1	0.15	0.5	1.

3.3.3 A Criterion for Evaluating the Seriation

As above, let us consider the seriation as defined on the set:

$$\tilde{\mathbb{I}} = (1, 2, ..., j, ..., n)$$

which indexes the row object set $\mathcal{O}$. Formally, this seriation is a permutation

$$\left(i_{(1)}, i_{(2)}, ..., i_{(j)}, ..., i_{(n)}\right)$$

of $\tilde{\mathbb{I}}$. This ordered sequence was denoted above in the following form:

$$\rho(\tilde{\mathbb{I}}) = (\rho(1), \rho(2), ..., \rho(j), , ..., \rho(n)) \tag{3.73}$$

(see Sect. 3.3.2). The quality of this permutation with respect to a Robinson structure of the following dissimilarity matrix

$$\rho(\mathcal{D}) = \{D\left(\rho(i), \rho(j)\right)|(i, j) \in \mathbb{I}^{[2]}\} \tag{3.74}$$

(see (3.53)) is assessed by using the criterion:

$$\mathbf{s}\left(\rho(\mathcal{D}), \mathbb{E}\right) = \sum_{(i,j)\in\mathbb{I}^{[2]}} D(\rho(i), \rho(j)).e(i, j) \tag{3.75}$$

(see (3.66)) where

$$\mathbb{E} = \{e(i, j)|(i, j) \in \mathbb{I}^{[2]}\} \tag{3.76}$$

(see (3.54)) where two options were considered for $e(i, j)$

$$e(i, j) = |i - j|^2 \text{ or } e(i, j) = |i - j|$$

(see (3.56) and (3.57)). In our development, we retain the easiest option, that is, $e(i, j = |i - j|$.

Now, given the permutation $\rho(\tilde{\mathbb{I}})$ (see (3.73)), obtained from an element $i_{(1)}$ chosen as an attraction pole, the coefficient $\mathbf{s}(\rho(\mathcal{D}), \mathbb{E})$ is nothing else than the raw association coefficient defined in Sect. 6.2.5.2 of Chap. 6 of [6], between the binary numerical valuations $\rho(\mathcal{D})$ and $\mathbb{E}$ (see (3.74) and (3.76)). A statistically normalized version of this coefficient is defined by

$$S(\rho(\mathcal{D}), \mathbb{E}) = \frac{\mathbf{s}(\rho(\mathcal{D}), \mathbb{E}) - \mathcal{E}[\mathbf{s}(\rho^\star(\mathcal{D}), \mathbb{E})]}{\sqrt{var[\mathbf{s}(\rho^\star(\mathcal{D}), \mathbb{E})]}} \tag{3.77}$$

where $\rho^\star$ is a random permutation in the set of all permutations on $\tilde{\mathbb{I}}$ provided with a uniform probability (see (6.2.77) and (6.2.86) of Sect. 6.2.5 of Chap. 6 of [6]). Clearly, the mathematical expectation and the variance in (3.77) are invariant whatever is the permutation ρ. We will specify them below in the case where the seriation concerns the set of column attributes of an incidence data table. Recall that this set was indexed by $\mathbb{J} = \{1, 2, ..., k, ..., p\}$.

Now, let us return to the framework of $\mathbb{I}$ endowed with a dissimilarity index D. And, let us imagine an algorithmic—optimizing step by step a local criterion— enabling us to produce a set of seriations on $\mathbb{E}$ that we have to evaluate each other. Consider the strict case for which each of the seriations induces a total order on $\mathbb{E}$, defining, thus, a permutation ρ on $\mathbb{I}$. The latter order is assessed by using the global criterion $S(\rho(\mathcal{D}), \mathbb{E})$ (see (3.77)). For the latter, the best seriation, defined by $\rho(\mathcal{D})$ is that for which $S(\rho(\mathcal{D}), \mathbb{E})$ is maximal. The statistical nature of the coefficient employed indicates how significant is the seriation obtained.

In Chap. 5, a family of combinatorial algorithms is proposed. Each of them produces a consistent set of seriations. These can be mutually compared by using the criterion $S(\rho(\mathcal{D}), \mathbb{E})$. Equally, other criteria can be employed (see Sect. 3.3.4).

The starting point in our approach is a data table crossing an object set $\mathcal{O}$ and an attribute set $\mathcal{A}$. We consider more particularly the basic and essential case of an incidence data table composed of zeros and ones. However, the approach can be extended to a large family of data table structures of formal type *Objects* × *Attributes*.

In these conditions, consider the case of an incidence data table. If the question is of applying the above technique as directly as possible to the set of objects (coded by $\mathbb{I}$), we will take as distance index that is defined by

$$D^2(i, j) = \sum_{1 \leq k \leq p} (x_i'^k - x_j'^k)^2 \tag{3.78}$$

$(i, j) \in \mathbb{I} \times \mathbb{I}$ (see (3.142) for the definition of $x_i'^k$).

Clearly, we can also work with the similarity table:

$$\mathcal{S} = \{S(i, j) | (i, j) \in \mathbb{I} \times \mathbb{I}\} \tag{3.79}$$

where

$$S(i, j) = \sum_{1 \leq k \leq p} x_i'^k . x_j'^k \tag{3.80}$$

(see (3.140) and (3.141)). Here the reader is asked to make explicit the formalism and the calculation of the expressions concerned in this case.

Now, the method proposed above in order to discover a seriation on the set of objects coded by $\mathbb{I}$ can be transposed to discover a seriation on the set of attributes (coded $\mathbb{J}$). In this case, the distance matrix becomes

$$\mathbf{d} = \{d(k, l)|(k, l) \in \mathbb{J}^{[2]}\} \tag{3.81}$$

where

$$d^2(k, l) = \sum_{1 \leq i \leq n} \left(x_i'^k - x_i'^l\right)^2 \tag{3.82}$$

$(k, l) \in \mathbb{J}^{[2]}$.

Without risk of ambiguity, we can still denote by ρ the permutation of $\mathbb{J}$ derived from a seriation on $\mathbb{J}$. The expression of the significance criterion of the seriation on $\mathbb{J}$, with respect to a Robinson structure of the permuted distance matrix, is put in the form (3.77). Although what follows is, conceptually, a repetition, let us rewrite here—in the context—this criterion:

$$S(\rho(\mathbf{d}), \mathbb{E}) = \frac{\mathbf{s}(\mathbf{d}, \mathbb{E}) - \mathcal{E}[\mathbf{s}(\rho^\star(\mathbf{d}), \mathbb{E})]}{\sqrt{var[\mathbf{s}(\rho^\star(\mathbf{d}), \mathbb{E})]}} \tag{3.83}$$

Notice that (3.77) is expressed with an initial labeling of $\mathbb{J}$ in which $\rho(k)$ occupies the kth row (resp., column) position. $\rho^\star$ is here a random permutation on $\tilde{\mathbb{J}} = (1, 2, ..., k, ..., p)$, taken in the set—provided with a uniform probability—of the $p!$ permutations on $\tilde{\mathbb{J}}$. Here, we will denote $\mathbb{E}$ directly in the form

$$\mathbb{E} = \{|k - l| \,|(k, l) \in \mathbb{J}^{[2]}\} \tag{3.84}$$

where, as expressed above, $\mathbb{J}^{[2]}$ is the set of ordered pairs of distinct elements of $\mathbb{J} = \{1, 2, ..., k, ..., p\}$. We have

$$\mathbf{s}(\mathbf{d}, \mathbb{E}) = \sum_{(k,l) \in \mathbb{J}^{[2]}} d(k, l).|k - l| \tag{3.85}$$

To be explicit we will give below the expressions of the expectation $\mathcal{E}$ and the variance var that are included in formula (3.83).

Therefore, by considering adaptation to $\mathbb{J}$ of methods established to seriate $\mathbb{I}$, we can, among a set of seriations—on the basis of the distance table $\mathbf{d}$ (see (3.81)) and the coefficient (3.83)—determine those (generally that) which maximizes the coefficient concerned.

The seriation of the set of column attributes (rites in archeology, species or genera in paleontology) is of fundamental interest. However, the expert can aim at the seriation of the row objects (tombs in archeology, sites in paleontology). Both seriations are related (see the mathematical definition of a σ form in Sect. 2.2). Under these conditions, the question arises if it is necessary to determine the seriation on $\mathbb{I}$ from that on $\mathbb{J}$.

In our strategy, we recommend to deduce the seriation on $\mathbb{I}$ from that on $\mathbb{J}$ (see Chap. 5, Sect. 5.4). The reason is given by the condition $C1P$ (equivalent to the condition of absence of Lazarus events) (see Sect. 2.2.1), for which the respective numbers of ones in the different columns are comparable. The latter condition makes the seriation on $\mathbb{J}$ more robust than that on $\mathbb{I}$.

3.3.3.1   The Components of the Calculation of the Evaluation Criterion

Relative to the criterion (3.83) the matter is to detail the elements of the calculation of the association coefficient between two symmetric binary valuations developed in Sect. 6.2.5 of Chap. 6 of [6]. To be as concrete as possible, we shall illustrate the components of this development on an example. To this end, we place ourselves in the context of Equation (3.77).

Imagine a set of column attributes comprising 16 elements. Thus, $\mathbb{J} = \{1, 2, ..., j, ..., 16\}$. The permutation identity $\rho_0(\mathbb{J})$ is the initial permutation

$$\rho_0(\mathbb{J}) = \tilde{\mathbb{J}} = (1, 2, 3, 4, 5, 6, 7, 8, 9, 10, 11, 12, 13, 14, 15, 16) \tag{3.86}$$

$\mathbb{J}$ is supposed endowed with the distance matrix (see (3.81)). We have

$$\mathbf{s}(\rho(\mathbf{d}), \mathbb{E}) = \sum_{(k,l)\in\mathbb{J}^{[2]}} d(\rho(k), \rho(l)).|k - l| \tag{3.87}$$

The evaluation of $\mathcal{E}[\mathbf{s}(\rho^\star(\mathbf{d}), \mathbb{E})]$ and $var[\mathbf{s}(\rho^\star(\mathbf{d}), \mathbb{E})]$ depends on Tables (3.81) and (3.84).

For the mathematical expectation (see (6.2.78) and (6.2.79) of Sect. 6.2.5 of Chap. 6 of [6]), we have

$$\mathcal{E}[\mathbf{s}(\rho^\star(\mathbf{d}), \mathbb{E})] = p^{[2]}.\mu_\mathbf{d}.\mu_\mathbb{E} \tag{3.88}$$

where $p^{[2]} = p(p-1)$ (16 × 15 for $p = 16$). $\mu_\mathbf{d}$ and $\mu_\mathbb{E}$ are the averages of the tables $\mathbf{d}$ and $\mathbb{E}$.
We have

$$\mu_\mathbf{d} = \frac{1}{p^{[2]}}. \sum_{(k,l)\in\mathbb{J}^{[2]}} d(k, l) \tag{3.89}$$

$$\mu_{\mathbb{E}} = \frac{1}{p^{[2]}} \cdot \sum_{(k,l) \in \mathbb{J}^{[2]}} |k - l| \tag{3.90}$$

The sum included in the preceding equation is equal to

$$\frac{(p-1).p.(p+1)}{3}$$

It is equal to 1360 for $p = 16$. Then, we have

$$\mu_{\mathbb{E}} = \frac{p+1}{3}$$

In these conditions, (3.88) becomes

$$\mathcal{E}[s(\rho^\star(\mathbf{d}), \mathbb{E})] = \frac{(p-1)p(p+1)}{3} \cdot \mu_{\mathbb{E}} \tag{3.91}$$

For the variance computing, we refer to the Mantel expressions [38]. These have been analyzed and worked in depth (see (6.2.83) to (6.2.85) of Sect. 6.2.5.3 of Chap. 6 of [6]). Moreover, let us point out that these expressions were interpreted in an original in [39].

Relative to a given table of numerical values as

$$\Delta = \{\delta(k, l) | (k, l) \in \mathbb{J}^{[2]}\} \tag{3.92}$$

the following quantities have to be calculated for each of the tables $\mathbf{d}$ and $\mathbb{E}$.

$$E_1 = \Big(\sum_{(k,l) \in \mathbb{J}^{[2]}} \delta(k, l) \Big)^2$$

$$E_2 = \sum_{1 \le k \le p} \Big(\sum_{l \in \mathbb{J} - \{k\}} \delta(k, l) \Big)^2$$

$$E_3 = \sum_{(k,l) \in \mathbb{J}^{[2]}} \delta(k, l)^2 \tag{3.93}$$

These quantities are denoted by A_1, A_2 and A_3, respectively, for the table $\mathbf{d}$ and B_1, B_2 and B_3 for the table $\mathbb{E}$. The calculation of B_1, B_2 and B_3 can be derived directly from combinatorial considerations. We obtain

$$B_1 = \left[\frac{(p-1)p(p+1)}{3} \right]^2$$

$$B_2 = \frac{1}{3}.(p-1)p(p+1)$$

$$B_3 = \frac{1}{6}.(p-1)p^2(p+1) \tag{3.94}$$

In these conditions, the Mantel expression of the variance is (see (6.2.85) of Sect. 6.2.5 of Chap. 6) of [6]:

$$
\begin{aligned}
var[s(\mathbf{d}, \rho^\star)] &= \frac{2}{p^{[2]}} A_3.B_3 \\
&+ \frac{4}{p^{[3]}}(A_2 - A_3).(B_2 - B_3) \\
&+ \frac{1}{p^{[4]}}(A_1 - 4A_2 + 2A_3).(B_1 - 4B_2 + 2B_3) \\
&- \frac{1}{p^{[2]^2}} A_1.B_1
\end{aligned}
\tag{3.95}
$$

where

$$
p^{[r]} = p(p-1)...(p-r+1)
$$

is the rth factorial exponent of p ($r = 1, 2, 3, ...$).

3.3.4 Usual Criteria for Evaluating a Seriation

To begin, let us recall two types of seriation evaluation criteria considered above. The first is more local than the second. The first one, conceptually, is defined at the level of the values observed in the data table $\mathcal{T}$ (see (2.2) in Chap. 2). The argument of the second type of criterion is a symmetrical similarity, dissimilarity or distance matrix between the elements of the set $\mathbf{E}$ to be organized according to a seriation. The first case of a data table structure is described in [33] as two-way two-mode data and the second, as two-way one-mode data. For the first type of criterion, we may refer to that used by Niermann (see (3.44), (3.45) and [26]). For the second type, we may refer to (3.83) considered above.

Now, we shall detail the local criteria we have examined in our experimental analysis. Most of them are mentioned in the literature [1, 33]. Nevertheless, some of them are completely new. These criteria do not necessarily take part in the algorithm of building the seriation but more essentially in its evaluation. This does not prevent us from following how a given algorithm optimizes, step by step a given criterion. In a way, each of the local criteria measures the concentration of the one value in the area of the data table, the latter being reorganized from joint permutations of rows and columns. These permutations are established in order to bring out the target structure, a seriation here. However, this structure can also be defined by clustering.

3.3.4.1 The Different Local Criteria

Column Lazarus Index: *Col-Laz* This index as the following, concern an incidence data table constituted by zeros and ones. It is defined as the number of Lazarus events. It is expressed in Chap. 2 around Equation (2.4). The absence of Lazarus event in a given column means that the set of entries in this column including the one value constitutes a connected interval of the row set. Thus, if i_l (resp., j_l) is the first row (resp., the last row) of the lth column ($1 \leq l \leq p$) including the one value, we have

$$(\forall i, i_l \leq i \leq j_l), x_i^l = 1$$

and

$$(\forall i, i < i_l \text{ or } i > j_l), x_i^l = 0 \text{ or not defined} \qquad (3.96)$$

The contribution of the l column to the Lazarus index is defined by the number of components equal to zero in this column, between two components equal to one. By denoting L_c^l this contribution, the column Lazarus index can be written as

$$Col\text{-}Laz = \sum_{1 \leq l \leq p} L_c^l \qquad (3.97)$$

Row Lazarus Index: *Row-Laz* This index is dual of the preceding one. The respective roles of rows and columns are interchanged. The absence of Lazarus event in a given row i means that the entries in this row including the one value constitutes a connected interval of the column set. Thus, if k_i (resp., l_i) is the first column (resp., the last column) of the row i, $1 \leq i \leq n$, we have

$$(\forall k, k_i \leq k \leq l_i), x_i^k = 1$$

and

$$(\forall k, k < k_i \text{ or } k > l_i), x_i^k = 0 \text{ or not defined} \qquad (3.98)$$

The contribution of the row i to the row Lazarus index is the number of components equal to zero in this row which are situated between two components equal to one. By designating L_r^i this contribution, the row Lazarus index can be written as

$$Row\text{-}Laz = \sum_{1 \leq i \leq n} L_r^i \qquad (3.99)$$

Column and Row Lazarus Index: *Col-Row-Laz* This index is defined by the sum of *Col-Laz* and *Row-Laz* (see (3.97) and (3.99)).

$$Col\text{-}Row\text{-}Laz = Col\text{-}Laz + Row\text{-}Laz \qquad (3.100)$$

Measure of Effectiveness 4: *ME4*   This criterion is defined in [40]. The value x_i^j of the cell (i, j) of the data table is compared to the respective contents of a neighborhood composed of a maximum four cells ($1 \leq i \leq n$, $1 \leq j \leq p$). Two cells $(i, j - 1)$ and $(i, j + 1)$ are located horizontally on the left and the right of the cell (i, j) concerned and two: $(i - 1, j)$ and $(i + 1, j)$, vertically, below and above the cell (i, j). Some of the cells $(i, j - 1)$, $(i, j + 1)$, $(i - 1, j)$ and $(i + 1, j)$ may not exist. In this case, they are not taken into account. More precisely, if (i, j) is placed next or just before one of the data table borders, we put

$$x_i^0 = x_i^{p+1} = x_0^j = x_{n+1}^j = 0 \tag{3.101}$$

With this convention, x_i^j is compared to the sum

$$y_i^j = x_i^{j-1} + x_i^{j+1} + x_{i-1}^j + x_{i+1}^j \tag{3.102}$$

Up to the multiplicative factor $1/2$, the *ME4* measure can be written in the form

$$ME4 = \sum_{1 \leq i \leq n} \sum_{1 \leq j \leq p} x_i^j \cdot y_i^j \tag{3.103}$$

where $x_i^j = 0$ or $x_i^j = 1$ and where y_i^j is between 0 and 4.

Notice that *ME4* (see (3.103)) can be interpreted as a raw similarity coefficient between the two numerical valuations

$$\{x_i^j | 1 \leq i \leq n, 1 \leq j \leq p\}$$
$$\text{and}$$
$$\{y_i^j | 1 \leq i \leq n, 1 \leq j \leq p\} \tag{3.104}$$

(see (5.2.63) of Sect. 5.2.2.2 of Chap. 5 of [6]).

In these conditions, statistical normalization of this coefficient can be considered (see Sect. 5.2.2.2 of Chap. 5 of [6]), leading to a correlative coefficient.

Measure of Effectiveness 8: *ME8*   The nature of this criterion is the same as the preceding one. However here, the neighborhood of a given cell (i, j), containing the value x_i^j is composed of a maximum of 8 cells adjacent to (i, j). This pattern of neighborhood is attributed to Moore [33]. Four of the cells are those considered above for *ME4*. Four additional cells are located diagonally with respect to (i, j): two $(i + 1, j - 1)$ and $(i - 1, j + 1)$ in the left to right direction and two others $(i + 1, j + 1)$ and $(i - 1, j - 1)$, in the right to left direction. So that, x_i^j has to be associated with

$$z_i^j = x_{i-1}^{j-1} + x_{i-1}^j + x_{i-1}^{j+1}$$
$$+ x_i^{j-1} + x_i^{j+1} + x_{i+1}^{j-1} + x_{i+1}^j + x_{i+1}^{j+1} \tag{3.105}$$

In the right member of the latter equation, if—for some of its elements—the lower subscript is equal to 0 ($i = 1$) or to $n + 1$ ($i = n$), the concerned value x_{i-1}^j which cannot exist is taken equal to 0. Similarly, if the higher subscript is equal to 0 ($j = 1$) or to $p + 1$ ($j = p$) the concerned value x_i^{j-1} or x_i^{j+1}, respectively, which cannot exist in the data table is taken equal to 0. We have

$$ME8 = \sum_{1 \leq i \leq n} \sum_{1 \leq j \leq p} x_i^j . z_i^j \tag{3.106}$$

As for *ME4*, we may notice that *ME8* corresponds to a raw similarity coefficient between two numerical valuations on the entry set of the data table, namely,

$$\{x_i^j | 1 \leq i \leq n, 1 \leq j \leq p\}$$
$$\text{and}$$
$$\{z_i^j | 1 \leq i \leq n, 1 \leq j \leq p\} \tag{3.107}$$

the latter being defined on

$$\{(i, j) | 1 \leq i \leq n, 1 \leq j \leq p\} \tag{3.108}$$

Correlation Coefficient According to Moore Neighboring: *Corr-8* As for *ME4*, different versions of a correlation coefficient can be expressed relatively to *ME8*, between the two valuations described in (3.107) (see Sect. 5.2.2.2 of Chap. 5 of [6]). We have indeed established and experienced a classic version of this coefficient. For this, additionally to *ME8*, the following calculations are needed:

$$mean(x) = \frac{1}{n \times p} \sum_{1 \leq i \leq n} \sum_{1 \leq j \leq p} x_i^j$$
$$mean(z) = \frac{1}{n \times p} \sum_{1 \leq i \leq n} \sum_{1 \leq j \leq p} z_i^j \tag{3.109}$$

$$mean(x^2) = \frac{1}{n \times p} \sum_{1 \leq i \leq n} \sum_{1 \leq j \leq p} (x_i^j)^2$$
$$mean(z^2) = \frac{1}{n \times p} \sum_{1 \leq i \leq n} \sum_{1 \leq j \leq p} (z_i^j)^2 \tag{3.110}$$

$$Corr\text{-}8(x, z) = \frac{ME8 - mean(x) \times mean(z)}{\sqrt{var(x) \times var(z)}} \tag{3.111}$$

where

$$var(x) = mean(x^2) - (mean(x))^2$$

$$\text{and}$$

$$var(z) = mean(z^2) - (mean(z))^2 \tag{3.112}$$

Clearly, we could also have considered a correlation coefficient associated with the raw similarity index *ME4* based on the neighborhood in the Neumann sense. This development is left for the reader.

Neumann Stress: *Neum-Strs* According to (3.101), the Neumann neighborhood of the cell (i, j) is defined by

$$\{(i, j - 1) \text{ or nil if } j = 1, (i, j + 1) \text{ or nil if } j = p,$$
$$(i - 1, j) \text{ or nil if } i = 1, (i + 1, j) \text{ or nil if } i = n\} \tag{3.113}$$

In these conditions, the contribution of the entry x_i^j to the Neumann stress is

$$Neum\text{-}Strs(x_i^j) = \sum_{k=max(1,j-1)}^{min(p,j+1)} (x_i^j - x_i^k)^2 + \sum_{l=max(1,i-1)}^{min(n,i+1)} (x_i^j - x_l^j)^2 \tag{3.114}$$

Then,

$$Neum\text{-}Strs = \sum_{1 \leq i \leq n} \sum_{1 \leq j \leq p} Neum\text{-}Strs(x_i^j) \tag{3.115}$$

Moore Stress: *Moor-Strs* According to (3.105), the Moore neighborhood of the cell (i,j) is defined as follows:

$$\{(i - 1, j - 1) \text{ or nil if } i = 1 \text{ or } j = 1, (i - 1, j) \text{ or nil if } i = 1,$$
$$(i - 1, j + 1) \text{ or nil if } i = 1 \text{ or } j = p,$$
$$(i, j - 1) \text{ or nil if } j = 1, (i, j + 1) \text{ or nil if } j = p,$$
$$(i + 1, j - 1) \text{ or nil if } i = n \text{ or } j = 1, (i + 1, j) \text{ or nil if } i = n,$$
$$(i + 1, j + 1) \text{ or nil if } i = n \text{ or } j = p. \} \tag{3.116}$$

In these conditions, the contribution of the entry x_i^j to the Moore Stress is

$$Moor\text{-}Strs(x_i^j) = \sum_{k=max(1,i-1)}^{min(n,i+1)} \sum_{l=max(1,j-1)}^{min(p,j+1)} (x_i^j - x_k^l)^2 \tag{3.117}$$

Then

$$Moor\text{-}Strs = \sum_{1 \leq i \leq n} \sum_{1 \leq j \leq p} Moor\text{-}Strs(x_i^j) \tag{3.118}$$

The *Neum-Strs* and *Moor-Strs* are clearly expressed in [33]. This is not the case for correlation coefficients based on *ME4* and *ME8*. These correspond to new propositions.

Also, the correlation coefficient associated with Deutsh and Martin method [25] is not made explicit in the two latter references. Let us detail it in the following.

The Deutsh and Martin Correlation Coefficient: *DM-Corr* In the **R** "Seriation" this coefficient is called *Cor-R* [41], we will name it *DM-Corr*. It is directly related to the Deutsh and Martin algorithm [25]. In the latter, at convergence, the values x_i^j of the data table are placed in such a way that, overall, including normalizations, we reach the stability of the respective values of

$$\sum_{1 \leq j \leq p} x_i^j \frac{j}{p} \tag{3.119}$$

$(1 \leq i \leq n)$ and

$$\sum_{1 \leq i \leq n} x_i^j \frac{i}{n} \tag{3.120}$$

$(1 \leq j \leq p)$

(3.119) defines the row locations where the respective rows have to be placed and (3.120), those where the respective columns have to be placed.

DM-Corr is a correlative measure of the quality of the joint assignment of the x_i^j $(1 \leq i \leq n, 1 \leq j \leq p)$ to the different rows and columns of the data table.

To begin, we introduce the following normalization factor:

$$T = \sum_{1 \leq i \leq n} \sum_{1 \leq j \leq p} x_i^j \tag{3.121}$$

Recall that the optimal assignation of the jth column is fixed from a normalization of (3.119). This is defined by $1/c_i$, where

$$c_i = \sum_{1 \leq j \leq p} x_i^j \tag{3.122}$$

Overall, the quality of the assignation of the different rows is measured by

$$\overline{Y} = \sum_{1 \leq i \leq n} \left(\sum_{1 \leq j \leq p} \frac{j}{p} . x_i^j \right) / T \tag{3.123}$$

Similarly, the optimal assignation of the jth column is fixed from a normalization of (3.120). This is defined by $1/c^j$ where

$$c^j_. = \sum_{1 \le i \le n} x^j_i \tag{3.124}$$

Overall, the quality of the assignation of the different columns is measured by

$$\overline{X} = \sum_{1 \le j \le p} \left(\sum_{1 \le i \le n} \frac{i}{n} . x^j_i \right) / T \tag{3.125}$$

In these conditions, the respective variances related to the rows and columns can be written as follows:

$$SY2 = \left[\sum_{1 \le i \le n} \left(\sum_{1 \le j \le p} \left(\frac{j}{p} - \overline{Y} \right)^2 x^j_i \right) \right] / (T-1) \tag{3.126}$$

and

$$SY2 = \left[\sum_{1 \le j \le p} \left(\sum_{1 \le i \le n} \left(\frac{i}{n} - \overline{X} \right)^2 x^j_i \right) \right] / (T-1) \tag{3.127}$$

The covariance in the assignation—globally calculated—of x^j_i at the ith row and the jth column can be put in the form

$$SXY = \left[\sum_{1 \le i \le n} \sum_{1 \le j \le p} \left(\frac{i}{n} - \overline{X} \right) . \left(\frac{j}{p} - \overline{Y} \right) x^j_i \right] / (T-1) \tag{3.128}$$

Finally,

$$DM\text{-}Corr = \frac{SXY}{\sqrt{SX2 \times SY2}} \tag{3.129}$$

3.4 Spectral Approaches

As pointed out above, most of the seriation methods address directly the problem of ordering a set of objects provided with a similarity index S. In these conditions, the data is defined by a similarity matrix S as that expressed in (3.59) that we take up again here

$$S = \{S(i,j) | (i,j) \in \mathbb{I}^{[2]}\} \tag{3.130}$$

where $\mathbb{I}$ indexes the object set.

The spectral methods are based on computing eigenvalues and eigenvectors of matrices which derive from S. In [35], the problem is posed in terms of totally ordering $\mathbb{I}$. That is—as already expressed in (3.63)—to find out a permutation $\tau(\mathbb{I})$

of $(1, 2, ..., i, ..., n)$, minimizing

$$\mathbf{u}(\mathcal{S}, \ll(\mathbb{I})) = \sum_{(i,j)\in\mathbb{I}^{[2]}} S(i,j).(\tau(i) - \tau(j))^2 \qquad (3.131)$$

Due to the discrete nature of a permutation, search in the permutation space on $\mathbb{I}$ is NP-hard [42]. Indeed, the latter space comprises $n!$ elements.

In order to avoid combinatorial optimization, a relaxation of the problem is proposed in [35]. In this, a numerical value x_i is substituted for the rank $\tau(i)$, $1 \leq i \leq n$. More precisely, the function

$$\mathbf{v}(x) = \sum_{(i,j)\in\mathbb{I}\times\mathbb{I}} S(i,j)(x_i - x_j)^2 \qquad (3.132)$$

is substituted for $\mathbf{u}(\tau)$. Scaling and normalization conditions must be taken into account for the vector of numerical values

$$\mathbf{x} = \left(x_1, x_2, ..., x_i, ..., x_n\right)^T$$

These are expressed as follows:

$$\sum_{i\in\mathbb{I}} x_i = 0 \text{ and } \sum_{i\in\mathbb{I}} x_i^2 = 1 \qquad (3.133)$$

The total ordering of the numerical components of the solution vector determines the seriation sought. Thus, the solution vector is returned to satisfy the monotony condition, that is to say: $x_i \leq x_{i+1}$ for all $1 \leq i < n$ or $x_i \geq x_{i+1}$ for all $1 \leq i < n$.

Now, let us recall the definition of the *Laplacian* $L_\mathcal{S}$ of a symmetric matrix $\mathcal{S}$. It can be written as follows:

$$L_\mathcal{S} = \Delta_\mathcal{S} - \mathcal{S} \qquad (3.134)$$

where $\Delta_\mathcal{S}$ is the diagonal matrix, whose ith element δ_i is the sum of the entries of the ith row of $\mathcal{S}$, that is,

$$(\forall i \in \mathbb{I}), \delta_i = \sum_{1\leq j\leq n} S(i,j) \qquad (3.135)$$

In these conditions, the minimizing problem of $\mathbf{v}(x)$ subject to the constraints (3.133) can be expressed in the following form:

$$min\{\mathbf{x}.L_\mathcal{S}.\mathbf{x}^T |^T \mathbf{x}.\mathbf{1} = 0, {}^T\mathbf{x}.\mathbf{x} = 1\} \qquad (3.136)$$

where $\mathbf{1}$ is the column vector whose all of the components are equal to 1.

The elements of the matrix $\mathcal{S}$ are assumed to be positive or null. The main result concerns the case for which the latter matrix is R irreducible (R as Robinson), that is to say, there does not exist a permutation π of $(1, 2, ..., i, ..., n)$ which reduces the matrix S to the form:

$$S^\pi = \begin{pmatrix} T & 0 \\ 0 & U \end{pmatrix}$$

Thereby, the minimal value defined in (3.136) is the lowest *no null* eigenvalue of L_S (see (3.134)). This eigenvalue is called *Fiedler* value in [35]. Note that the vector **1** is an eigenvector associated with a null eigenvalue. It realizes the minimization defined in (3.136).

In real cases, we may expect that the irreducibility condition is not always satisfied. And then, it is necessary to manage situations for which the matrix S is reducible to a sequence of diagonal blocks (there are two blocks suggested above). A technique based on the observation of repetitions in the components of the Fielder eigenvector enables the blocks of the matrix S to be identified. Thus, in the case, where the latter matrix can be brought back by permutation to the Robinsonnian form.

An observation. Compared to the technique used in the non-irreducibility case, it is questionable whether a clustering technique would be equally suitable (see Sect. 3.2).

The formalization and the main result obtained in [35] are exploited and developed in a specific way in [43]. The objective consists of building an original solution to the very known problem of finding out a total ordering on a set of items (objects) from all ordered pairwise comparisons between these items. It is indeed a very classical problem which appeared in social sciences and voting theory (de Borda 1781, de Condorcet 1785). More recently, we can cite Zermelo (1929) [44]. The most cited historical reference is probably (Kendall and Smith 1940) [45].

Whereas the similarity matrix S is not specified in [35], in [43] there is a specific adaptation set for the (i, j) element of the matrix S, more precisely,

$$S(i,j) = \sum_{1 \leq k \leq n} s_{ij}^k \qquad (3.137)$$

where for all (i, j, k), $1 \leq i, j, k \leq n$

$$s_{ij}^k = 1 \text{ if } (i > k \text{ and } j > k) \text{ or } (i < k \text{ and } j < k)$$

$$s_{ij}^k = \frac{1}{2} \text{ if } k = min(i, j)$$

$$s_{ij}^k = 0 \text{ if } min(i, j) < k < max(i, j)$$

$$s_{ij}^k = \frac{-1}{2} \text{ if } k = max(i, j) \qquad (3.138)$$

In these conditions and in the case of a total ordering of $\mathbb{I}$, for which i occupies the position of rank i, it can be seen that

$$S(i,j) = n - 1 - |i - j| \qquad (3.139)$$

Thereby, the matrix $\mathcal{S}$ is qualified as a strict **R** matrix [43].

3.4.1 Some Methodological Points and Extensions

To begin, let us adapt the [35] approach in our framework. It is simply a matter of knowing what we have to substitute for formula (3.132) expressing $\mathbf{v}(x)$. The basic data in our approach is an incidence table where the attributes are Boolean and recall that we focus on ordering the attribute set $\mathcal{A}$. Let us continue to refer to this data structure, but according to the mentioned reference, consider the problem of the seriation of the object set $\mathcal{O}$. For this purpose, it is needed to specify the matrix

$$\mathcal{S} = \{S(i,j)|(i,j) \in \mathbb{I} \times \mathbb{I}\} \tag{3.140}$$

For this specification, we refer to (2.12) of Sect. 2.3.2 of Chap. 2, and then we propose

$$\left(\forall (i,j) \in \mathbb{I} \times \mathbb{I}\right), S(i,j) = \sum_{1 \le k \le p} x_i'^k . x_j'^k \tag{3.141}$$

where, recall

$$\left(\forall (i,l), 1 \le i \le n, 1 \le l \le p,\right), x_i'^l = \frac{x_i^l - \eta_l}{\sqrt{\eta_l} . \sqrt{n}} \longrightarrow x_i'^l \tag{3.142}$$

where η_l is the proportion of objects for which the Boolean attribute a^l is *TRUE*.

The index $S(i,j)$ is of the same nature as that underlying the normalized version of principal component analysis.

Now, if the matter is the seriation of the attribute set $\mathcal{A}$, the coefficient to consider is that (2.13) of Sect. 2.3.2 of Chap. 2. Let us recall its expression:

$$S(k,l) = \sum_{1 \le i \le n} x_i'^k . x_i'^l \tag{3.143}$$

This coefficient corresponds to a correlation coefficient between the attributes a^k and a^l.

Each of both indices (see (3.141) and (3.143)) is open to be extremely generalized whatever are the respective natures of the attributes, that is, whatever are the respective structures on the object set $\mathcal{O}$, induced by these attributes.

In the case of ordering the attribute set $\mathcal{A}$, for which (3.143) has to be generalized, all the descriptive attributes have to be of the same type (for example, *numerical* or *ordinal categorical*). Therefore, if α and β are two attributes from $\mathcal{A}$, the diagram of Fig. 6.1 of Chap. 6 of [6] is followed in order to establish a statistically normalized association coefficient in the following form:

$$Q(\alpha, \beta) = \frac{s(\alpha, \beta) - \mathbb{E}(s(\alpha^\star, \beta^\star))}{\sqrt{var(s(\alpha^\star, \beta^\star))}} \tag{3.144}$$

where $s(\alpha, \beta)$ is a raw association coefficient between α and β and where $\alpha^\star$ and $\beta^\star$ are two independent random attributes, respectively, associated with α and β (see Sect. 6.2.1 of Chap. 6 of [6]). $\mathbb{E}$ and var designate the expectation and variance. The similarity coefficient we propose between two elements in the similarity matrix $\mathcal{S}$ as it operates in the [35] method has the same form as (6.2.5) of Chap. 6 of [6], that is:

$$R(\alpha, \beta) = \frac{Q(\alpha, \beta)}{\sqrt{Q(\alpha, \alpha).Q(\beta, \beta)}} \tag{3.145}$$

This coefficient has been widely experienced in [39] (see also Sect. 11.5 of Chap. 11 of [6]).

Now, for the seriation of the object set $\mathcal{O}$ in the case of heterogeneous description by attributes of all types, a large extension of the index (3.141) is provided in Sect. 7.2 of Chap. 7 of [6]. The index concerned is denoted by Q^g (see (7.2.4) and (7.2.5)) of the cited reference and more particularly (see Sect. 7.2.6).

A contingency table (see (5.3.1) of Sect. 5.3.2 of Chap. 5 of [6]) is a perfectly symmetrical statistical structure. Rows and columns play the same role in relation to each other. Rows (resp., columns) are defined by an exclusive and exhaustive set of categories. As in (5.3.1) we denote by $\mathbb{I} = \{1, 2, ..., i, ..., I\}$ (resp., $\mathbb{J} = \{1, 2, ..., i, ..., J\}$) the set of subscripts indexing the row categories (resp., the column categories). Generally, by convention and convenience, $\mathbb{J}$ is interpreted as associated with an attribute set and $\mathbb{I}$, as associated with an object set. In this way, all the row classes can each include a single object. Therefore, in the case of $\mathbb{J}$ seriation according to [35] view, the (k, l) element of the similarity matrix $\mathcal{S}$ is defined by (5.3.5) of the cited reference, $1 \leq k, l \leq J$. Now, in the case of $\mathbb{I}$ seriation, the similarity matrix $\mathcal{S}$ can be specified by either

$$\mathcal{S} = \{Cor_0(i, i')|(i, i') \in \mathbb{I} \times \mathbb{I}\} \tag{3.146}$$

or

$$\mathcal{S} = \{Cor_g(i, i')|(i, i') \in \mathbb{I} \times \mathbb{I}\} \tag{3.147}$$

(see (7.2.39) and (7.2.40) of the cited reference)

In the spectral method of correspondence analysis [46, 47], the structure of the data table is viewed as that of a contingency table. This method is practiced with specific coding and interpretation rules in the case of Boolean or numerical data [48]. In these conditions and as already expressed just above, the data table is reduced to the following relative frequency distribution:

$$f_{\mathbb{I} \times \mathbb{J}} = \{f_{ij}|(i, j) \in \mathbb{I} \times \mathbb{J}\} \tag{3.148}$$

crossing on the basis of a same object set $\mathcal{O}$ two exhaustive and exclusive sets of categories (see (5.3.1) and (5.3.2) of Sect. 5.3.2 of Chap. 5 of [6]). f_{ij} is the proportion of objects which possess jointly the categories i and j, $(i, j) \in \mathbb{I} \times \mathbb{J}$.

Let us here address the seriation of $\mathbb{J}$ as an attribute set (see just above). This, because we focus in our approach the seriation of the attribute set. Anyway, seriation of $\mathbb{I}$ can be got in the same fashion or in a dual fashion by using (7.2.40). Compared to the formalism developed in [35], the (j, k) element of the similarity matrix S of which we have to determine the eigenvalues and the associated eigenvectors is

$$\sum_{1 \leq i \leq n} \frac{f_{ij}.f_{ik}}{p_{i.}.\sqrt{p_{.k}.p_{.j}}} \qquad (3.149)$$

Lebart and Fenelon [49] and Sect. 7.2.8 of Chap. 7 of [6].
Here, the eigenvector associated with the eigenvalue 0 is

$$u^0 = (\sqrt{p_{.1}}, \sqrt{p_{.2}}, ..., \sqrt{p_{.j}}, ..., \sqrt{p_{.J}}) \qquad (3.150)$$

Now, consider in the set of the eigenvectors orthogonal to the vector u^0 the first two vectors u^1 and u^2 associated with the two *largest* eigenvalues of S. These will sustain the representation of $\mathbb{J}$ in a geometrical plane. This representation will reveal the seriation of $\mathbb{J}$ when this ordinal structure underlies the data table. We will experiment it in the case of simulated and real data (see Chap. 4).

As a matter of fact, correspondence analysis [46, 47] is known for emphasizing seriations described as "Guttman Effect" in [50]. In the geometrical plane provided, the points representing the column attributes (resp., the row objects) appear arranged according to a parabolic curve. This is to compare with the shape of the "Horse-shoe" curve obtained by the Kendall method (see Sect. 3.5.2 and [51, 52]).

As it will expressed in the following Sect. 3.5.1, correspondence analysis can be derived in a specific way from principal component analysis, this in order to analyze rows through columns (resp., columns through rows). Both methods can be used to reveal a seriation. This will be experimented in Chap. 4. One approach we consider consists of ordering the attribute set (indexed by $\mathbb{J}$), followed by specific algorithm which deduces the seriation on the object set from that attribute set (see Chap. 5, Sect. 5.2).

As we can see, there is an essential conceptual difference between the latter representations and the treatment applied in [35] for which the Fiedler value is the *smallest* non-trivial (no null) eigenvalue value. Technically, this might be due to the fact that the opposite $-S$ of the similarity matrix appears in the Laplacian of S. On the other hand, in the spectral analysis [35], a single eigenvector associated with a single eigenvalue intervenes to reveal a seriation. This is not the case for geometric spectral methods based—for more precise discrimination—on an ordered pair of eigenvectors associated with two consecutive eigenvalues.

3.5   Methods Using a Planar Geometrical Representation

3.5.1   Introduction: The Most Classical Methods

We have just mentioned above the application of the factorial methods of correspondence and principal component analyses for bringing out seriations. A seriation
appears from the planar geometrical representation of the column set (resp., the row
set) of the data table. It is in Chap. 4 that geometrical representation methods will be
experienced. These are comprised of the two preceding factorial methods, the attraction pole based on similarities and the multidimensional scaling horse-shoe methods.
Simulated data according to σ forms and real data will be considered. In these experiments, formally, these methods address an incidence data table describing a set of
objects by a set of Boolean attributes.

We have already described the planar geometrical version of the attraction pole
method in Sect. 2.4.5 of Chap. 2, for representing the set of column attributes. Technically, it is the simplest geometrical representation method. It does not require to
compute eigenvalues and the associated eigenvectors. By considering the philosophy of this approach and the similarity coefficients used, we have compared it with
principal component analysis (see hereafter and the end of Sect. 2.4.5 of Chap. 2).
The simplicity of the attraction pole method makes that it can be methodologically
situated between the visual and the factorial methods.

To begin—even if we have to come back to it in the next chapter (Chap. 4)—
let us outline the principles of the most classical factorial analysis methods in data
analysis. These are principal component analysis and correspondence analysis (see
[49], Chap. 6 of [6], [47, 50] and [48]). The first of both methods addresses a numerical rectangular table, defining description of an object set in rows by a numerical
attribute set in columns. A factor is defined as a linear combination of the descriptive attributes, these being interpreted as linear coordinate forms in the geometrical
space $\mathbb{R}^p$ where p denotes the attribute number. The importance of a given factor is
measured by its *variance*. This can be interpreted as a part of the inertia of a cloud
of points in $\mathbb{R}^p$ associated with the set of the row objects. A *factor* is viewed as pro-
jective measure on an axis in $\mathbb{R}^p$, the latter being called *factorial axis*. The reduced
geometrical representation, mostly considered, is sustained by the two first factorial
axes, corresponding to the largest variances. Column attribute representation is associated with row object representation. Each of the attributes is positioned according
to its correlation coefficients with the two first factors.

Now, let us give a general intuitive idea of correspondence analysis. This will
be reconsidered in Chap. 4. The technical nature of this method is exactly the same
as that of principal component analysis. It is addressed to two dual analyses of the
elements—as they are arranged—in a contingency table. In the first analysis, rows are
interpreted as representing objects and columns, as representing numerical attributes.
In the second analysis, columns are interpreted as representing objects, and rows,
as representing attributes. The first analysis (resp., the second analysis) is that of
rows through the columns (resp., of columns through the rows). For the first analysis

(resp., the second analysis), with each row (resp., column) is associated its *profile* through the column sequence (resp., through the row sequence). For a given analysis, geometric representation of the set playing the role of the object set is considered. Specific *weighting* on the cloud of points obtained and a specific *metric* providing the representation space are adopted. These make that the cloud total inertia is exactly the Chi-square of the contingency table [53]. As just indicated, the first factorial analysis leads to the row representation and the second one to the column representation.

In fact, the incidence data table, we are interested in, is composed of zero and one Boolean values, where the zero indicates *absence* and one, *presence*. Applying principal component analysis to this data table assumes to liken Boolean values to numerical values. This, to some extent, can be justified (see Chap. 4 of [6]).

Now, in relation to applying correspondence analysis, the incidence data table is likened to a contingency table. For this, we can imagine the set of rows as corresponding to the finest partition for which each class is a singleton. However, more generally, the column set does not define a partition on the set of objects. It defines a family of subsets of the latter set, these not being mutually disjoint. Nevertheless, the practical results of this method as those of principal component analysis have proven to be very interesting and significant.

The horse-shoe method proposed by Kendall that we will present briefly in the next section is indeed a planar geometric representation method. It differs from factorial analysis methods in that it is not based on the search for eigenvalues and associated eigenvectors. It is also distinguished in that the data retained for the optimization of the representation is a *preordonance* (see Sect. 4.2.2 of Chap. 2 of [6]) on the set to order. Usually, this preordonance is associated with a similarity or a dissimilarity numerical index on the set concerned. More precisely, we apply the principles of the Shepard-Kruskal approach [9, 54] and the *MDSCAL* algorithm of Kruskal [9]. In the latter, we optimize a distance function called "stress" between the given preordonance and that reconstructed one in the planar geometrical representation. Let us point out that this function is a numerical one. It does not handle directly—in a combinatorial way—the preordonances, respectively, associated with the similarity distributions of the clouds of points in the original space and that in the planar geometrical representation.

3.5.2 The Horse-Shoe Method of Kendall

Kendall begins by justifying the merits of his method by considering a general seriation proceeding within combinatorial optimization (see Sect. 3.3). The set to be seriated is the object set, coded by $\mathbb{I} = \{1, 2, ..., i, ..., n\}$. Three important stages are considered in the latter type of approach:

1. Choice of a similarity index on the object set leading to a similarity matrix Σ.
2. Construct a measure of discrepancy $D(\Sigma)$ between the ordinal structure of Σ and that obtained by joint permutation of its rows and columns which gives it

a Robinson form (recall that this condition expresses that for any three subset $\{h, i, l\}$ of the row set, for which the order is $h < i < l$, we have $s(h, i) \geq s(h, l)$ and $s(i, l) \geq s(h, l)$ (see (3.69) and (3.70)).

3. Specify a permutation algorithm on $\mathbb{I}$ where the initial permutation is chosen or randomly determined.

In the third stage, an algorithm step—that we have to specify—consists of substituting the current permutation on $\mathbb{I}$ by that which results from the biggest decrease of $D(\Sigma)$ and that by applying the latter permutation jointly on the rows and columns of Σ. Generally, the last step leads to a local minimum of $D(\Sigma)$.

Kendall [8] believes that the choices involved in this type of process have an arbitrary character. These are the choices of $\Sigma, D(\Sigma)$ and the progression algorithm from one permutation of $\mathbb{I}$ to the next one.

However, most often the Σ matrix imposes itself naturally and in any case, the final result is relatively stable compared to a reasonable choice of the index of similarity on the set of objects.

Let us note here that in combinatorial optimization methods, we consider rather a matrix $\mathcal{D}$ of dissimilarity or distance on the set $\mathcal{O}$ of the objects (3.51). Clearly, we can associate with a similarity matrix a dissimilarity one. In the case of our concern, for an incidence data table where the attributes are Boolean, we will pass from the following similarity index:

$$S(i, j) = \sum_{1 \leq k \leq p} x_i'^k . x_j'^k \tag{3.151}$$

to the distance index $D(i, j)$, using the formula

$$D^2(i, j) = \sum_{1 \leq k \leq p} (x_i'^k - x_j'^k)^2 \tag{3.152}$$

where (see (2.102) in Chap. 2), with the following notations already seen

$$x_i'^k = \frac{x_i^k - \eta_k}{\sqrt{\eta_k . \sqrt{n}}}$$

$1 \leq i, j \leq n$

In fact, a dissimilarity matrix Δ on $\mathbb{I}$, such that

$$\Delta = \{\delta(i, j) | (i, j) \in \mathbb{I} \times \mathbb{I}\} \tag{3.153}$$

defines a complete valued graph on $\mathbb{I}$. In this graph, the search of a Hamiltonian path, comprising $n - 1$ edges of minimal length results, from a certain point of view, an optimal seriation with respect to the dissimilarities between adjacent vertices. The length of the latter path gets in the following form:

$$\sum_{1 \le i \le n-1} \delta(i, i+1) \tag{3.154}$$

Reference [34, 55].

Nevertheless, the algorithmic computation complexity to search such a path is NP.

In our combinatorial approach, expressed in Chap. 5, we will propose a family of methods statistically justified and of linear complexity with respect to the size of the dissimilarity matrix (resp., of the similarity matrix), once the latter acquired. More particularly, we will be set in the case where the dissimilarity (resp., similarity) function is derived from an incidence data table. These methods are able to address directly the seriation of the object set. However, it has been proved experimentally that more relevant results can be obtained when we derive the seriation of the object set from that of the attribute set.

The Horse-shoe method is very different in its principle. It addresses the problem of direct search of a seriation on the set of objects, represented by the rows of the data table [8, 56, 57]. Ordering the set the column attributes is not concerned. As mentioned above, this method is principally based on the integration and behavior analysis of the Kruskal algorithm [9]. This algorithm called *MDSCAL* (MultiDimensional SCALing) attempts to carry out an idea of Shepard [54] which can be stated as follows:

"Given a set of n points in a geometrical space of dimension p, determine an Euclidean representation of the latter set in a space with a small number of dimensions in such a way to approximate as well as possible the *inequalities* between the interpoints distances calculated in the initial geometrical space".

In the problem posed, the set considered of the n points is that associated with the rows of the incidence data table, each row defining an object description. Therefore, the representation space is $\{0, 1\}^p$. This is immersed in the geometrical space $\mathbb{R}^p$. The mutual distances between the different points in the initial representation space are given by the classical Euclidean metric. In these conditions, the inner product

$$s(h, i) = \sum_{1 \le j \le p} x_h^j . x_i^j \tag{3.155}$$

defines the proximity between two object points (representing two tombs in the archeological example). $s(h, i)$ is the number of attributes (rites in archeology) which are possessed by the two objects h and i, $1 \le h < i \le n$. (3.155) is in fact the raw similarity index between h and i, for a Boolean description (see (4.2.3) of Sect. 4.2.1.2 of Chap. 4 of [6]).

As in [8], let Σ designate a similarity matrix. This can be defined by

$$\mathcal{S} = \{s(h, i) | 1 \le h, i \le n\} \tag{3.156}$$

or deduced from the latter (see (3.157)).

A dispersion criterion $D(\Sigma)$ is defined in order to evaluate the distortion of the Σ similarity structure with respect to the situation where the Robinson condition is satisfied. Recall that this condition expresses that for any three subset $\{h, i, l\}$ of the row set, for which the order is $h < i < l$, we have $s(h, i) \geq s(h, l)$ and $s(i, l) \geq s(h, l)$.

If the seriation problem is posed in terms of direct seeking for a row permutation in order to minimize $D(\Sigma)$, we find oneself back in the situation of combinatorial optimization as considered above (see Sect. 3.3).

Here, we are concerned with the application of the *MDSCAL* algorithm. Its argument is a similarity matrix Σ, this matrix being either S or $S \cdot S$ (see (3.157)). In the latter case, the (h, i) entry can be written as follows:

$$(S \cdot S)_{hi} = \sum_{1 \leq l \leq n} w_l . min\big(s(h, l), s(i, l)\big) \qquad (3.157)$$

where w_l is a weight defined generally by the expert and translating the importance to accord for the association with the lth object, $1 \leq l \leq n$.

As already mentioned above the version used of the *MDSCAL* algorithm is that for which the preordonance associated with the similarity index (see Sect. 4.2.2 of Chap. 4 of [6]) is approximated. The total preorder on the set of object pairs, defined by the preordonance structure, is coded with the "mean rank" function.

The representation considered is a planar geometrical one. It can be sought directly. Another alternative, previously considered for obtaining it, consists, first, of applying *MDSCAL* in order to determine the representation of the object points in a three-dimensional geometrical space. The cloud of points thus obtained is projected on the plane sustained by the two first eigenvectors associated with the 3×3 inertia matrix derived from the cloud of points.

Whatever the alternative chosen, suppose that there exists a column permutation followed by a row permutation giving rise to a σ form of the incidence data table. By applying *MDSCAL* one obtains—after a sufficient number of iterations—an arrangement of the row points in the representation plane, in the form of a horse-shoe curve. The order of these points on this curve is—up to a reversion—the row order according to a σ form of the incidence data table. This order enables—in archeology—the tomb chronological order to be retrieved.

Kendall extends his analysis to the case for which the value denoted by x_i^j—in the cell at the intersection of the object row i and the column attribute j—is defined by the number of times where the Boolean attribute j (expressing a rite in archeology) is present in the object i (expressing a tomb in archeology). In these conditions, in each column j of a σ form, the ordered sequence of the x_i^j values is either

1. Increasing till a maximum value and then decreasing.
2. Increasing till a maximum value.
3. Decreasing till a minimum value.

The preceding method based on the *MDSCAL* algorithm addresses directly the seriation problem of the object set $\mathcal{O}$. As emphasized several times above (see, for example, Sect. 3.1), our approach consists of starting by ordering the set of column attributes. In these conditions, the ordering of the row set is derived by using an algorithmic procedure (see Sect. 5.2 of Chap. 5).

3.6 Combinatorial and Algorithmic Approaches of the Pole Attraction Method

As just mentioned above, a new family of combinatorial algorithms will be developed in Chap. 5. In these, the attraction can be performed from a subset of a few elements. Although the computational complexity of the algorithms defined is linear with respect to the similarity or dissimilarity (proximity) matrix of the set concerned, the optimization matter is combinatorial (see Sect. 3.3).

Here, we shall describe more fully this type of approach in a very simple case where the attraction is carried out from a single element of the set to be seriated. Two fundamental principles govern this. The first one is to determine the attraction pole which drives by successive linking the total order that best corresponds to the similarities (resp., dissimilarities) of the set to seriate.

The second fundamental point lies in the method of determining the element that must follow the last element already associated with the seriation under construction. More precisely, $x_{(1)}$ being the chosen as attraction pole, if $x_{(1)}, x_{(2)}, ...,$ and $x_{(l)}$ are the first extracted ordered elements, it is a matter of determining how we obtain the $(l + 1)$th element $x_{(l+1)}, 2 \le l \le m - 1$, where m indicates the size of the set to order (seriate).

Now, consider, as in Sect. 3.3, the case where the seriation is to be sought directly on the set of objects $\mathcal{O}$. We assume that the latter is encoded with $\mathbb{I} = \{1, 2, ..., i, ..., n\}$ and provided with a dissimilarity index D (see (3.51)). To begin, let us illustrate the second fundamental point mentioned above. For this, designate by $(i_{(1)}, i_{(2)}, ..., i_{(l)})$ the ordered sequence of the l first elements of the seriation. $i_{(l+1)}$ is defined as belonging to $\mathbb{I} - \mathbb{I}_l$, such that

$$i_{(l+1)} = Arg\left\{min\{D(i_l, i)|i \in \mathbb{I} - \mathbb{I}_l\}\right\} \tag{3.158}$$

where $\mathbb{I}_l = \{i_{(1)}, i_{(2)}, ..., i_{(l)}\}$.

In words, $i_{(l+1)}$ is the element of the remaining subset (not yet seriated) whose dissimilarity with the last seriated element $i_{(l)}$ is minimal. In the rare cases where there exists more than a unique solution to (3.158), we may imagine exploring the different solutions.

In fact, each of the $\mathcal{O}$ elements will be evaluated as the first element $i_{(1)}$ of the ordering. So that, the rule (3.158) will generate n seriations. The best one according to the value of the criterion calculated for $i_{(n-1)}$ is retained.

The experimental results of this technique will not be reported in Chap. 5. In the latter, we have retained more efficient methods. For these, the proximity notion of

$i_{(l+1)}$ to the ordered sequence $(i_{(1)}, i_{(2)}, ..., i_{(l)})$ is more finely established. Moreover, the attraction pole concept is extended and can correspond to a fairly small subset.

With regard to the methods that can be associated with the one expressed above (see (3.158)), we can consider finding the shortest Hamiltonian path traversing n vertices of a given geometric configuration. In these conditions, the question of how to establish the relation between such a path and Robinson's form of the similarity matrix (resp., dissimilarity matrix) has to be examined [55]. A particular point concerns the origin of the Hamiltonian path and the initial element of the seriation.

More directly, we may refer to the *Bond Energy Algorithm* (**BEA**) whose principle has to be compared with the algorithm defined here (see [58], [40] and Chap. 5 of [59]). Suppose a description data table of an object set $\mathcal{O}$ by a Boolean attribute set $\mathcal{A}$ (two-way two-mode data) and consider the ordering problem of $\mathcal{O}$. In the **BEA**, an element of $\mathcal{O}$ is chosen randomly and placed first as corresponding to $i_{(1)}$. Now, relative to the established sequence $(i_{(1)}, i_{(2)}, ..., i_{(l)})$ $(l < n)$, the next step consists of placing an $(l + 1)$th element, $1 \leq l \leq n - 2$, providing from the remaining row set, the latter comprising $(n - l)$ elements. This element is chosen in such a way that its placement leads to the highest value of *ME4* (see (3.103)) by taking into account the already established state of the data table. The choice for its placement can be before the first element $i_{(1)}$, between two consecutive elements $i_{(k)}$ and $i_{(k+1)}$ $(1 \leq k \leq l - 1)$, or after the last element $i_{(l)}$. In these conditions, $(n - l) \times (l + 1)$ alternatives have to be tested before choosing the element to be placed. Notice that by this way the criterion value (*ME4*) for positioning the $(l + 1)$th element does not only depend on the already placed elements.

To finish, let us mention that **BEA** is generally described to order the set of column attributes first and next, the row objects. In fact, this is what we do. However, the philosophy we follow is very different. Consider that ordering column attributes corresponds to ordering descriptive variables of an object set. This problem has focused component factorial analysis for many years. Conceptually, a descriptive data table (two-way two-mode data) is not symmetrical. Otherwise, suppose having a similarity (resp., dissimilarity) matrix on a set **E** to be ordered in such a way that, overall, two consecutive elements are as near as possible. The principle of our approach is to begin by considering an exhaustive set of ways to start the seriation. For a given way, the seriation is obtained by successive chainings, using the similarity (resp., dissimilarity) matrix. This approach will be extensively developed in Chap. 5.

References

1. D.L. Carlson, *Quantitative Methods in Archeology Using R* (Cambridge University, 2017)
2. F. Marcotorchino, Block seriation problems: a unified approach. Appl. Stochast. Models Data Anal. **3**, 73–91 (1987)
3. I.-C. Lerman, *Les bases de la classification automatique* (Gauthier-Villars, 1970)
4. J.A. Hartigan, *Clustering Algorithms* (Wiley, 1975)
5. B. Ganter, R. Wille, *Formal Concept Analysis* (Springer, 1999)

6. I.C. Lerman, *Foundations and Methods in Combinatorial and Statistical Data Analysis and Clustering* (Springer, 2016)
7. J. Bertin, *La Graphique et le traitement graphique de l'Information* (Flammarion, 1977)
8. D.G. Kendall, Seriation from abundance matrices, in *Mathematics in Archaeological and Historical Sciences*, ed. by D.G. Kendall F.R. Hodson, P. Tautu (Chicago, Aldine-Atherton, 1971), pp. 214–252
9. J.B. Kruskal, Multi-dimensional scaling. Psychometrika **29**, 1–27, 28–12 (1964)
10. J. Bertin, Traitements graphiques et mathématiques. différence fondamentale et complémentarité. Mathématiques et Sciences Humaines **80**, 60–71 (1980)
11. I. Liiv, Seriation and matrix reordering methods: an overview. Wiley InterScience (www.interscience.wiley.com), (https://doi.org/10.1002), 70–91, Jan. 2010
12. G. Gruvaeus, H. Wainer, Two additions to hierarchical cluster analysis. Br. J. Math. Stat. Psychol. **25**, 200–206, 200-206
13. G. Brossier, Algorithmes d'ordonnancement des hiérarchies binaires et propriétés. Revue de Statistique Appliquée **32**, 65–79 (1984)
14. G. Brossier, *Problèmes de représentation par des arbres, Thèse de Doctorat ès Sciences*. Ph.D. thesis, Université de Rennes 2, June 1986
15. G. Brossier, Classification hiérarchique à partir de matrices carrées non symétriques. Statistique et Analyse des Données **7**, 22–40 (1982)
16. D. Vicari, Classification of asymmetric proximity data. J. Classif. **13**, 386–420 (2014)
17. F. Murtagh, *Data Science Foundation; Geometry and Topology of Complex Hierarchic Systems and Big Data Analytics* (Chapman and Hall, 2018)
18. V. Elisséeff, Possibilités du scalogramme dans l'étude des bronzes chinois. Mathématiques et Sciences Humaines **11**, 1–10 (1964)
19. V. Elisséeff, De l'application des propriétésdu scalogramme à l'étude des objets. *Calcul et formalisation dans les sciences de l'homme* (1968), pp. 107–120
20. V. Elisséeff, Donnés de classement fournies par les scalogrammes privilégiés, in *Archéologie et Calculateurs*, ed. by M. Borillo J.-C. Gardin (Centre National de la Recherche Scientifique, 1970), pp. 177–180
21. I.C. Lerman, *Classification et analyse ordinale des données*. Dunod and http://www.brclasssoc.org.uk/books/index.html (1981)
22. I.C. Lerman, Analyse du phénomène de la "sériation" à partir d'un tableau d'incidence. Mathématiques et Sciences Humaines **38**, 39–57 (1972)
23. I.C. Lerman, H. Leredde, La méthode des pôles d'attraction, in *Analyse des Données et Informatique*, ed. by E. Diday et al. (IRIA, 1977), pp. 37–49
24. H. Leredde, *La méthode des pôles d'attraction, La méthode des pôles d'agrégation*. Ph.D. thesis, Université de Paris 6, Oct. 1979
25. S.B. Deutsch, J.J. Martin, An ordering algorithm for analysis of data arrays. Oper. Res. **19**(6), 1350–1362 (1971)
26. S. Niermann, Optimizing the ordering of tables with evolutionary computation. Am. Stat. Assoc. **59**, 41–46 (2005)
27. J.H. Holland, *Adaptation in Natural and Artificial Systems*. Ph.D. thesis, University of Michigan (1975)
28. D. Goldberg, *Genetic Algorithms in Search, Optimization and Machine Learning* (Addison Wesley, 1989)
29. T. Vallée, M. Yildizoglu, Présentation des algorithmes génétiques et de leurs applications en économie. Revue d' Économie Politique 711–745 (2004)
30. R.F. Ngouënet, *Analyse Géométrique des Données de Dissimilarité par la Multidimensional Scaling : Une Approche Basée sur les Algorithmes Génétiques*. Ph.D. thesis, Université de Rennes 1, Dec. 1995
31. J.B. Kruskal, M. Wish, *Multidimensional Scaling* (Sage Publications, 1984)
32. I.C. Lerman, R.F. Ngouenet, Algorithmes génétiques séquentiels et prallèles pour une représentation affine des proximités. Research Report 2570, IRISA-INRIA, Jan. 1995

33. M. Hahsler, K. Hornik, C. Buchta, Getting things in order: an introduction to the R package seriation. J. Stat. Softw. **25**(3), 1–34 (2008)
34. G. Caraux, S. Pinloche, Permutmatrix: a graphical environment to arrange gene expression profiles in optimal linear order. Bioinformatics **21**(7), 1280–1281 (2005)
35. J.E. Atkins, E.G. Boman, B. Hendrickson, A spectral algorithm for seriation and the consecutive ones problem. Siam J. Comput. **28**(1), 297–310 (1998)
36. P. Arabie, L.J. Hubert, Combinatorial data analysis. Annu. Rev. Psychol. **43**, 169–203 (1992)
37. W.S. Robinson, A method for chronologically ordering archaeological deposits. Am. Antiq. **16**, 293–301 (1951)
38. N. Mantel, Detection of disease clustering and a generalized approach. Cancer Res. **27**(2), 209–220 (1967)
39. M. Ouali-allah, *Analyse en préordonnance des données qualitatives. Application aux données numériques et symboliques*. Ph.D. thesis, Université de Rennes 1, décembre 1991
40. W.T. Mccormick, P.J. Schweitzer, T.W. White, Problem decomposition and data reorganization by a clustering technique. Oper. Res. **20**(5), 993–1009 (1972)
41. M. Hahsler, C. Buchta, K. Hornik, Infrastructure for ordering objects using seriation. Technical, Package 'Seriation', June 1919
42. A. George, A. Pothen, An analysis of spectral envelope-reduction via quadratic assignment problems. SIAM J. Matrix Anal. Appl. **18**, 706–732 (1997)
43. F. Fogel, A. d'Aspremont, M. Vojnovic, Spectral ranking using seriation. J. Mach. Learn. Res. **17**, 1–43 (2016)
44. E. Zermelo, Die berechnunug der turnier-ergebnisse als ein maximumproblem der wahrschein-lichkeitsrechnung. Mathematishe Zeitschrift **29**(1), 436–460 (1929)
45. M.G. Kendall, B.B. Smith, On the method of paired comparisons. Biometrika **31**(3-4), 324–345 (1940)
46. M. Hill, Correspondence analysis: a neglected multivariate method. Appl. Stat. **23**, 340–354 (1974)
47. J.-P. Benzécri, *L'analyse des données, tome II* (Dunod, 1973)
48. M. Greenacre, *Correspondence Analysis in Practice* (Chapman & Hall, 2007)
49. L. Lebart, J-P. Fenelon, *Statistique et Informatique Appliquées* (Dunod, 1973)
50. J-P. Benzécri, *Pratique de l'Analyse des Données* (Dunod, 1980)
51. J. de Leeuw, A horseshoe for multidimensional scaling. *Department of Statistics UCLA* (2008), pp. 1–21
52. P. Diaconis, S. Goel, S. Holmes, Horseshoes in multidimentionnal scaling and local kernel methods. Ann. Appl. Stat. **2**(3), 777–807 (2008)
53. H.O. Lancaster, *The Chi-squared Distribution* (Wiley, 1969)
54. R.N. Shepard, The analysis of proximities: multidimensional scaling with an unknown distance function. Psychometrika **27**, 219–246 (1962)
55. L.J. Hubert, Some applications of graph theory and related nonmetric techniques to problems of approximate seriation: the case of symmetric proximity measures. J. Math. Stat. Psychol. **27**, 133–153 (1974)
56. D.G. Kendall, Some problems and methods in statistical archeology. World Archaeol. **1**(1), 68–76 (1969)
57. D.G. Kendall, Incidence matrices, interval graphs and seriation in archeology. Pac. J. Math. **28**(3), 565–570 (1969)
58. W.T. Mccormick, S.B. Deutsh, J.J. Martin, P.J. Schweitzer, Identification of data structures and relationships by matrix reordering techniques. Research Report TR P-512, Arlington, Institute for Defense Analyses (1969)
59. C. Ray, *Distributed Database Systems* (Dorling Kindersley, 2009)

Chapter 4
Comparing Geometrical and Ordinal Seriation Methods in Formal and Real Cases

4.1 Introduction

In this chapter, methods mathematically described in Chaps. 2 and 3 will be mutually compared experimentally with respect to the seriation problem. We shall be interested in revealing a column ordering and an associated row ordering in an incidence data table composed of zeros and ones, as that of Fig. 2.1, or more specifically, as that figured in the diagram 2.4. Simulated data and real data will be treated. In the first case, simulations will be performed according to parallelogrammatic σ forms (see Chap. 2, Sect. 2.2.1). Two real data are tested. They have given much to the literature in archeological seriation. These are corresponding to Merovingian belt buckles-plates or buckles-plates and to Münsingen-Rain celtic necropolis. To test the behavior of the proposed methods in the cases of different types of a σ form, the experimental strategy is defined as follows:

1. Start with a σ form of an incidence data table;
2. Permute randomly the rows and the columns of this table.

In these conditions, the matter consists of reconstructing the σ form by using a seriation method on the columns and rows of the permuted incidence data table.

We shall limit the experimental analysis to the parallelogram σ forms. In addition, the only cases considered are those for which the parameter α (see Definition 1 in Chap. 2) is equal to $1/n$. Recall that in this case the bottom translation of the vertical segment loaded with values equal to 1 from the $(l-1)$th column to the lth one, is equal to one. On the other hand, the extent of the latter segment is $e = n.\eta$ (see (2.52)–(2.57) in Chap. 2). In fact, the comparisons to be carried out in this case are very representative of the respective behaviors of the different seriation methods to compare. Otherwise, these comparisons are, indeed, laborious enough. The reader interested may consider cases for which the extent of the mentioned translation is greater than one. As an example, the configuration $n = 21$, $p = 10$, $e = 3$ and $\alpha = 2/21$, can be experienced.

© The Author(s), under exclusive license to Springer Nature Switzerland AG 2022
I. C. Lerman and H. Leredde, *Seriation in Combinatorial and Statistical Data Analysis*, Advanced Information and Knowledge Processing,
https://doi.org/10.1007/978-3-030-92694-6_4

Spectral methods are largely involved in seriation literature. They provide a geometrical representation in which the ordinal structure sought is numerically interpreted (see Sects. 3.4 and 3.5 of Chap. 3). In this framework, the more classical methods as Principal Component Analysis and Correspondence Analysis have to be considered first. Similarly, the method described in [1] could have been taken into account. Surprisingly, the authors concerned by the latter article do not situate their approach with respect to the most classical methods. It is of importance to note that the spectral methods require the calculation of the eigenvalues and the associated eigenvectors of the similarity matrix on the set of the elements to order.

This chapter focuses on the use of the geometric representation given by data analysis methods in order to highlight a seriation structure behind an incidence data table. As just expressed, we shall observe the behavior of different methods on simulated and real data. Two basic and very classical methods are Principal Component Analysis and Correspondence Analysis. Attraction Poles based on Similarities is a new method (see Sect. 2.4.5, Chap. 2). It needs to be compared with more classical methods. Let us indicate that, conceptually, in a way, the Attraction Poles based on Similarities method can be situated between visual and spectral methods (see Sects. 3.2 and 3.4 of Chap. 3). These methods will be designated by APS for Attraction Poles based on Similarities, PCA for Principal Component Analysis and CA for Correspondence Analysis. Equally and very importantly, the "horse-shoe" method of Kendall, that we designate by MDS-HS, has to be considered. It played a significant part in seriation methods. It is based on the multidimensional scaling approach, developed by Kruskal and collaborators [2].

We shall consider three configurations for the σ parallelogram form, in the most important case where the latter is constituted by a single block—in other terms—is connex. For the first one, the vertical segment loaded with values identically equal to one of the first column and that of the last column, have a non-empty intersection with respect to the row set. The σ form defined by this type of the incidence data table is called *strongly chained* (see on the left of Fig. 4.1 in Sect. 4.4.2).

For a *weakly chained* parallelogram σ form, the intersection of the vertical segment—loaded with values identically equal to one—starting the first column and that ending the last column, is empty.

We shall consider two weakly chained σ forms. The first will be called "*moderately weakly chained*" or, more briefly, "*moderately chained*" and the second one, "*weakly chained*" (see Fig. 4.1 in Sect. 4.4.2.1).

In addition, we shall consider cases for which the incidence data table can reveal two or three disjoint blocks, corresponding each to a parallelogram σ form (see Fig. 4.2 in Sect. 4.4.2.1).

Results in the case of simulated σ forms will be given in Sect. 4.4.2. Real data (Merovingian buckles-plates and Münsingen-Rain celtic necropole) is in Sect. 4.4.3.

In the following two sections (see Sects. 4.2 and 4.3), we will detail the general context underlying the development of this chapter, more particularly, the made use statistical and algorithmic methods or techniques, will be given.

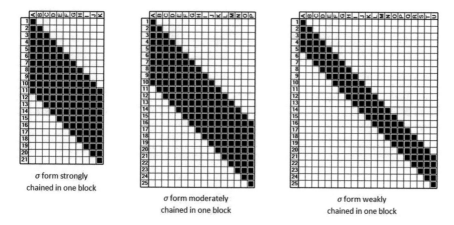

σ form strongly
chained in one block

σ form moderately
chained in one block

σ form weakly
chained in one block

Fig. 4.1 Three typical σ forms with one block

4.2 Methods

4.2.1 Similarities and Distances

To begin, let us specify that in the experiments carried out reference is systematically made to the data table

$$T = \{x_{ij} | 1 \le i \le n, 1 \le j \le p\} \tag{4.1}$$

where columns represent attributes (variables) and rows, objects (observations) and where $x_{ij} = 1$ (resp., $x_{ij} = 0$) if the jth Boolean attribute is *TRUE* (resp., *FALSE*) on the ith observation, $1 \le i \le n, 1 \le j \le p$. It is relative to this data table that similarities and distances are calculated in an Euclidean way (see Eq. (2.12)). In this, the metric concerned can be associated with that of the PCA method. Another alternative is to establish similarities and distances according the Chi-square metric which underlies the CA method. Let us give here as an indication the correlation coefficient and the distance between two column attributes according to Chi-square metric [3–5]. For this, start by associating with this data table that of relative frequencies

$$F = \{f_{ij} | 1 \le i \le n, 1 \le j \le p\} \tag{4.2}$$

where

$$f_{ij} = \frac{x_{ij}}{N} \text{ and where } N = \sum_{1 \le i \le n} \sum_{1 \le j \le p} x_{ij} \tag{4.3}$$

Now, introduce

$$x_{i.} = \sum_{1 \le j \le p} x_{ij} \text{ and } x_{.j} = \sum_{1 \le i \le n} x_{ij} \qquad (4.4)$$

$1 \le i \le n, 1 \le j \le p$

In these conditions, we consider the representation of $\mathbb{I} = \{1, 2, \ldots, i, \ldots, n\}$ through $\mathbb{J} = \{1, 2, \ldots, j, \ldots, p\}$. With each element i of $\mathbb{I}$ its profile through $\mathbb{J}$ is associated, that is,

$$f_{\mathbb{J}}^i = (f_1^i, f_2^i, \ldots, f_j^i, \ldots, f_p^i)$$
$$\text{where}$$
$$f_j^i = \frac{x_{ij}}{x_{i.}} \qquad (4.5)$$

Therefore, the following cloud of points in the geometrical space $\mathbb{R}^p$, is introduced

$$\mathcal{N}(\mathbb{I}) = \{(f_{\mathbb{J}}^i, p_{i.}) | i \in \mathbb{I}\}$$
$$\text{where}$$
$$p_{i.} = \sum_{1 \le j \le p} f_{ij} \qquad (4.6)$$

$p_{i.}$ is the proportion of one value in the ith row, $1 \le i \le n$.

In this spatial geometrical representation, each of the p attributes is represented by a component of $\mathbb{R}^p$. Besides, $\mathbb{R}^p$ is provided by the following diagonal metric:

$$q(e_j, e_k) = \frac{\delta_{jk}}{p_{.j}} \qquad (4.7)$$

where $\{e_j | 1 \le j \le p\}$ is the canonical base of $\mathbb{R}^p$, where δ_{jk} is the Kronecker symbol ($\delta_{jk} = 1$ if $j = k$ and 0 if not, $1 \le j, k \le p$) and where

$$p_{.j} = \sum_{1 \le i \le n} f_{ij} \qquad (4.8)$$

$1 \le j \le p$

In these conditions, the similarity index (or association coefficient) between the attributes a^j and a^k is the correlation coefficient between the linear forms $e_j^{\star}$ and $e_k^{\star}$, defined by the respective projections (components) on the vectors e_j and e_k. This gives

$$\rho(j, k) = \frac{\sum_{i \in \mathbb{I}} p_{i.}(f_j^i - p_{.j})(f_k^i - p_{.k})}{\sqrt{(\sum_{i \in \mathbb{I}} p_{i.}(f_j^i - p_{.j})^2) \times (\sum_{i \in \mathbb{I}} p_{i.}(f_k^i - p_{.k})^2)}} \qquad (4.9)$$

This formula is that (5.3.5) of [6], p. 237. In the latter book, formula (5.3.6) gives a simplified expression.

Up to a dual transformation, the correlation coefficient on $\mathbb{I}$ is given by equation (7.2.41), p. 351 of the latter reference.

Now, to define the distance index between j and k, we must consider the dual cloud

$$\mathcal{N}(\mathbb{J}) = \{(f_{\mathbb{I}}^j, p_{.j}) | j \in \mathbb{J} \tag{4.10}$$

In these conditions, we have

$$d_\chi^2(j, k) = \sum_{i \in \mathbb{I}} \frac{1}{p_{i.}} \times (f_i^j - f_i^k)^2 \tag{4.11}$$

$1 \le j, k \le p$

The nature of the data table and the association coefficients, similarities or distances between columns (representing Boolean attributes) or rows (representing objects or observations) will be completely specified in our development. This is not the case in the [1] proposed treatment, the latter being reported in Sect. 3.4 of Chap. 3.

4.2.2 Multidimensional Data Analysis

As already mentioned in the above introduction section (see Sect. 4.1, Chap. 4), the methods used are those of multidimensional data analysis. Let us recall that these are Principal Component Analysis (PCA), Correspondence Analysis (CA), the new method of Attraction Poles based on Similarities (APS) and the "horse-shoe" (MDS-HS) method. Each of these methods enables planar geometrical representation to be carried out, and this, as well for the column attributes as for the row objects. Nevertheless, referring to the PCA foundation method, we consider first the column attribute representation. Indeed, the linear form defined by the dimension relative to a given factorial axis, is a weighted mean of the different descriptive attributes interpreted as numerical variables [3–5, 7]. Moreover, considering the C1P condition [1], intuitively, more robustness can be expected in first seeking an order on the column set, the latter corresponding to the variables.

The case of CA method is very specific. This because the rows play in relation to the columns the same role as that the columns in relation to the rows. Nevertheless, we will start by showing the column attributes representation before that of row objects.

In the PCA, CA and APS methods, the coordinate axes of the geometrical representation are directly defined from the column attributes. Related to this point let us recall that in the CA method as in PCA method, component factorial analysis is applied for a transformed data table and according the Chi-square metric.

This type of development is not applied in the case of the MDS-HS method. In fact, the data of the latter is a set E provided with a dissimilarity or a distance index. Global numerical optimization is envisaged in order to produce an Euclidean representation, respecting as much as possible the dissimilarities or the distances. This method is more oriented towards the representation of row objects than towards column attributes. However, given the general orientation of our writing, we will start by performing the representation of the set of column attributes before that of row objects.

A total order is induced on the attribute set from the planar geometric representation of the column attributes. Several techniques can be proposed for this objective. Dually, a total order on the set of row objects is induced from the planar geometric representation of the row objects. Crossing both orders makes it possible to restitute the seriation σ form of the incidence data table in the case where this form is sufficiently chained. However, the whole approach becomes difficult to apply in the case of too weakly linked σ forms. It is then necessary to resort to algorithms of combinatorial nature (see Chap. 5).

4.3   Techniques and Algorithms

Following the end of the preceding section (see Sect. 4.2.2, Chap. 4), our strategy will be based on a planar geometrical representation of column attributes as given by multidimensional data analysis. The representation plane is referred to two coordinate axes, thus introducing two components. These axes are orthogonal in each of the PCA, CA and MDS-HS methods, but generally not in the APS method. Indeed, in the latter, the two axes are led, respectively, by the first two attraction poles.

As already indicated previously, we focus on the geometrical representation of the variables (of the attributes). Conceptually, as said, this is in relation to the founding PCA method, for which a factor defining a component on an axis, is directly expressed by a linear combination of the initial variables. Also, note that in the definition of a σ form, the order defined by the seriation structure refers more directly (more strictly, more explicitly) to the arrangement of the columns than to that of the rows (see Sect. 2.2, Chap. 2).

However, as mentioned in the previous section, the ordering of the rows can be achieved in the same way as that of the columns. That is to say from a planar geometric representation of the set of rows. This representation is of factorial nature for PCA, CA and APS.

Fundamentally, recall that the MDS-HS method is not a factorial one. It is based on monotonic regression relative to a dissimilarity or to a distance matrix between the elements of the set to be represented. Besides, this matrix can be provided from a rank matrix associated with a preordonance on the set concerned (see Chap. 4 of [6]). As said, MDS-HS is mainly employed for the representation of the row objects. In fact, we will show both attribute (variable) and object (observation) representations in applying this approach (see Sect. 4.4).

The σ forms (see Sect. 4.4.2.3, Figs. 4.1 and 4.2) constitute idealized seriation structures associated with mathematical models (see Sect. 2.2 of Chap. 2). Application on these data analysis methods allows us to realize the respective behaviors of these methods in controlled and differentiated cases. For each of these models and for a given analysis method—on column attributes or row objects—the representative points in the geometric plane are generally arranged regularly according to a concave or a convex curve (see in Sect. 4.4).

To fix ideas, suppose these representative points as corresponding to the columns (or to the rows) of an incidence data table provided from an instance of a σ form. Given the regularity in the evolution of these points, a polygonal line can be drawn traversing all of the points in an angular direction (see in Sect. 4.4). Each segment of this polygonal line joins two consecutive and near points. This technique is also practiced in analyzing real data (see Sect. 4.4.2).

Specifically, we start by identifying a starting point of the polygonal line and a circular direction in order to describe the line concerned. This direction can be clockwise or counterclockwise. Our visual perception and our intuition are used for this start point location and the direction chosen. Nevertheless, an automatic determination could contribute to these choices. Let S_0 be the origin point of the polygonal line. This point can be located in one of the four quadrants of the representation plane referred to two coordinate axes: either the fourth or the first or the second and or the third. Now consider the vector $\overrightarrow{OS_0}$, where O is the intersection point of the two coordinate axes. This vector determines two angular sectors in the quadrant in which it is located. They are in either side of this vector.

Very generally, the density of the points represented in one of the sectors is appreciably greater than in the other one. The position of the most heavily loaded sector indicates the direction of the scan (clockwise or anticlockwise) for describing the polygonal line. Examples will be given in the following Sect. 4.4.

More formally, denote by $\overrightarrow{OX}$ and $\overrightarrow{OY}$ the two coordinate axes which are directly on either side of $\overrightarrow{OS_0}$. As just said, O is the origin point at the intersection of the two axes. X designates a point at infinity on one of the axes and Y, on the other one. Without generality restriction, we suppose the angular sector $\widehat{S_0OX}$ more loaded in representative points than that $\widehat{S_0OY}$. This indicates that the angular direction of the circular scan of all representative points should be performed from $\overrightarrow{OS_0}$ to $\overrightarrow{OX}$.

Let $(P_1, P_2, \ldots, P_l, \ldots, P_m)$ be the sequence of the representative points following the circular order considered. These points can be associated with the columns or (exclusively) with the rows of the incidence data table. Now, set an initial angle as follows:

$$StrtAngl = \widehat{XOS_0} \tag{4.12}$$

In these conditions, consider the following sequence of polar angles:

$$\{Angl(l)|1 \leq l \leq m\} \tag{4.13}$$

where

$$Angl(l) = \widehat{XOP_l} \qquad (4.14)$$

$1 \leq l \leq m$.

Then, we operate the angular translation of value

$$TrsAngl = 2\pi - StrtAngl \qquad (4.15)$$

on the sequence (4.13) of angular values, and this, in order to obtain the angular sequence

$$\{AnglTrs(l) = (Angl(l) + TrsAngl \text{ modulo } 2\pi)\} \qquad (4.16)$$

Therefore, if the scan has to be performed in the anticlockwise direction the sequence $(P_1, P_2, \ldots, P_l, \ldots, P_m)$ is described according to increasing values of $AnglTrs(l)$, $1 \leq l \leq m$. And, in the opposite case for which the scan has to be executed clockwise, the sequence of points is described according to decreasing values of $AnglTrs(l)$, $1 \leq l \leq m$.

By joining the final reached point to the initial one S_0, we obtain a closed polygonal curve. Generally, the latter comprises in its interior a very substantial zone of the rectangle circumscribing all the representative points. Thereby, we realize the crucial importance of the choice of the initial point S_0. Examples of this choice are given in Sect. 4.4.2.3, below.

As expressed multiple times, a given result is defined by a pair of total orders, the first on the column set and the second on the row set of the incidence data table. Clearly, it is understood that the columns represent attributes (variables) and rows, objects (observations). The respective permutations of the columns and rows, according to these two orders, give to the data matrix—composed of zeros and ones—its seriation shape.

Each method and the associated technique produce a result. The latter can be assessed by each of the criteria defined in Sect. 3.3.4 of Chap. 3. These are noted as follows:

Col-Laz, Row-Laz, Col-Row-Laz, ME4, ME8, Corr-8, DM-Corr,

Moor-Strs and *Neum-Strs*

These are local criteria. Their respective evaluations are directly obtained from the reorganized data table composed of zeros and ones. A global criterion whose argument is a dissimilarity or distance matrix between the elements of the set concerned (columns or rows) is given by Eq. (3.77) in Chap. 3. We will note it *Lerm-Crit*.

It is of importance to realize that an optimized value of a criterion has not an absolute meaning, this value depends on the method and the associated technique applied. More precisely, the statistical nature of the criterion and the algorithm used are fundamental. Indeed, each algorithm carries within, its own semantics. Thus, given the same evaluation criterion Γ and two algorithmic procedures A and B, faced with data, if B optimizes Γ more than does A, the two results must nevertheless be regarded with the same importance. Indeed, the data scientist may find different interests in each of the two results. Thus, Hodson replaced the optimized result of Kendall's MDS-HS algorithm, by another result—close but different—corresponding more to archeological reality [8]. Moreover, the objective in our development is not so much to determine the *best* method, but, to describe in a relative way the different methods and to situate them in relation to each other.

In the framework of our approach, we will propose relatively to each type of method a technique of combinatorial nature enabling us to determine the seriation resulting from the order on the columns and from the order on the rows, the latter being derived from the order on the columns. Two algorithms intervene in this chapter. As just said, given their combinatorial nature, we will express them in Chap. 5. Their respective acronyms are OP2-S ("Ordinal Progressive Proximity with respect to a Single element") (see Sect. 5.3.1) and ROFCO ("Row Order From Column Order") (see Sect. 5.2.3). For more details, the reader is asked to refer to these references. Their common argument is a planar geometric representation of the set of column attributes. Relative to this representation, OP2-S uses a single element (point) which has an extreme position in this representation. From the latter, by successive chaining an ordering on the attribute set is established. ROFCO uses a complete ordering on the elements of the attribute set. This ordering can be recognized directly from the geometrical representation or deduced from applying OP2-S algorithm. ROFCO resumes the order on the rows. Crossing these two orders (on the columns and on the rows) enables the seriation form of the data table to appear.

Now, given a result by one of the methods in terms of seriation (i.e. pair of orderings on the column set and on the row set, resp.), this can be considered as an initial state to another algorithm. Here, it may be about of an evolutionary algorithm like that proposed by Niermann (see Sect. 3.2.6 of Chap. 3 [9]).

4.4 Results in Processing Data

4.4.1 Preamble

In this section, different simulated and real examples will be developed. This allows us to realize how geometric representation methods constitute a possible approach and a contribution to the seriation problem. As already expressed in the introduction section (see Sect. 4.1), the geometric methods applied are first, the methods of factor analyzes: Principal Component Analysis (PCA) and Correspondence Analysis (CA). Next, it is the method of Pole Attraction based on Similarities (APS) that is applied. Likewise, it is the "horse-shoe" method using multidimensional scaling approach (MDS-HS) which is taken into account. Finally, we consider a successive chaining algorithm (OP2-S), initiated by an origin element located in a given geometric representation of the column attributes by one the methods mentioned. This last algorithm is combinatorial in nature and therefore, described in Chap. 5.

In fact, we will resume here various aspects of the introduction (see Sect. 4.1) to which we refer. As indicated, we will, on the one hand, consider the processing of incidence data according to different versions of the sigma form (see 4.4.2). Besides, we will consider the processing of real data which occupied a very important place in archeology. These are, on the one hand, Merovingian buckles-plates and on the other hand, Münsingen-Rain celtic necropolis (see 4.4.3).

Let us recall that to test an analysis method on a data incidence table, we start by randomly permuting all of its rows and columns.

4.4.2 Simulated Data According to σ Forms Models

4.4.2.1 The Data Description

We shall consider three configurations for the σ parallelogram form, in the most important case where the latter is constituted by a single block—in other terms—is connected.

For the first one, the vertical segment loaded with values equal to one of the first column and that of the last column, have a non-empty intersection with respect to the row set. This configuration will be illustrated by the example of the left in Fig. 4.1, where $n = 21$, $p = 11$ and $e = 11$ (recall that n, p and e are the row number, the column number and the number of one entry per column, resp.). The σ form defined by this type of the incidence data table is called *strongly chained*.

For a *weakly chained* parallelogram σ form, the intersection of the vertical segment—loaded with values equal to one—starting the first column and that ending the last column, is empty. The σ form is as weakly chained as the difference $c(p) - e$ is large, the definition of the function $c(j)$, $1 \le j \le p$, being given at the beginning of Sect. 2.2.2 of Chap. 2. The "distance" in terms of row number between the last entry of loaded with 1 segment of the first column and the first entry of the such a segment in the last column is $c(p) - e - 1$.

We shall consider two weakly chained σ forms. For the first one (see in the middle of Fig. 4.1), $n = 25$, $p = 16$ and $e = 10$. In this case, $c(p) - e = 16 - 10 = 6$. For the second one (see on the right of Fig. 4.2), $n = 25$, $p = 21$ and $e = 5$. In the latter case, we have $c(p) - e = 21 - 5 = 16$. The first form will be called "*moderately weakly chained*" or, more briefly "*moderately chained*" and the second one, "*weakly chained*".

The mathematical results concern strongly chained seriation forms (see Theorems 1 and 2 in Chap. 2) and also, moderately chained seriation forms. In both cases, the planar geometrical representation will enables us to find out the ordering induced by the seriation. In this, the specification of the first element (the first point) is crucial (see Sect. 4.3).

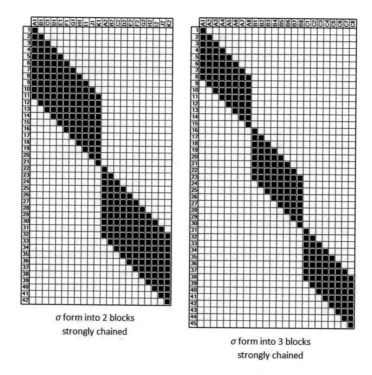

σ form into 2 blocks
strongly chained

σ form into 3 blocks
strongly chained

Fig. 4.2   Two or three disjoint blocks according to a σ form

In addition, we shall consider cases for which the incidence data table can reveal two or three disjoint blocks, corresponding each to a parallelogram σ form.

In the case of two blocks, the parameters of the incidence data table seriated are $n = 42$, $p = 22$ and $e = 11$. The restriction of the whole incidence data table to one of these two blocks is such as $l = 21$, $q = 11$ and $e = 11$, where l and q are the row and column numbers, resp.,. Moreover, the second block—from left to right—is deduced from the first one (on the left) by a translation (see on the left of Fig. 4.2).

We have experienced a second configuration into two blocks, where each of them is defined by a σ parallelogram form. In the latter the second block on the right, is deduced from the first one, on the left, by a translation followed by an homothety. For this configuration comprising two disjoint blocks strongly chained, $n = 38$, $p = 18$ and $e = 11$. The parameters of the incidence table restriction to the block on the left (resp., on the right) are $l = 21$, $q = 11$ and $e = 11$ (resp., $l = 17$, $q = 7$ and $e = 11$) where, as above, l and q are, resp., the row and column numbers concerned. The results obtained were very comparable to those, in the case of two identical blocks. Therefore we will be satisfied with the latter case.

For the configuration into three equal blocks (see the figure at the extreme right of Fig. 4.2), $n = 45$, $p = 24$ and $e = 8$, each parallelogram block is strongly chained and can be inscribed in a rectangle comprising 15 rows and 8 columns. The three identical blocks are disjoint. The second rectangle (resp., the third rectangle) can be deduced from the first one (resp., the second one) by a vector translation bringing the entry $(1, A1)$ to $(16, B1)$ (resp., the entry $(16, B1)$ to $(31, C1)$).

In the presentation of these results, we consider the moderately chained form as playing a pivotal role. It is for this reason that we present it first in sufficient detail. Afterwards, we will present more briefly

- σ form strongly chained in one block;
- σ form weakly chained in one block;
- σ form into two blocks, each strongly chained;
- σ form into three blocks, each strongly chained.

4.4.2.2 σ Form Moderately Chained in One Block

We consider below, on the one hand, on the left, the original form of the seriation on the data incidence table and, on the other hand, this same table after performing a joint permutation of the rows and columns. These two tables are accompanied by their respective tables of local criteria as defined in Sect. 3.3.4.1 of Chap. 3. It is the table as on the right of Fig. 4.3—that is to say whose rows and columns are first permuted at random—which is processed. The data analysis methods concerned are mentioned above: PCA, CA, APS and MDS-HS. Since the argument of the same method is a matrix of similarities or dissimilarities between the elements concerned, if in the course of the algorithm equal values are not encountered, the results are the same regardless of the random permutation of columns and rows considered. And then, we retrieve the starting seriation. This type of treatment gives confidence in the results obtained.

If in the progress of the algorithm equal values of the similarity are not encountered, we find the starting seriation.

Results when applying PCA Method

These two forms, semi-circular and circular (see Fig. 4.4), reflect the existence of a seriation underlying the data. A first approach for bringing out this seriation by an automatic procedure consists of projecting columns and rows on the first factorial axis concerned.

In accordance with our technique allowing us to recover an order on the rows from an order on the columns with the ROFCO algorithm (see Sect. 5.2.3 of Chap. 5), we can restore the target seriation by crossing the two orders. We can observe that, starting from the only order on the columns provided by the PCA analysis, we can obtain a noticeable improvement in the order on the rows (see on the right Fig. 4.5).

Also, we can see that by following these more or less circular shapes, we can propose slightly different orders by following in one direction or in the opposite one, the circular sequence of the elements. This provides jointly an order on the rows and an order on the columns.

Here, we propose to replace this purely visual observation by an automated process. However, the starting point of the latter cannot be automated. It consists in

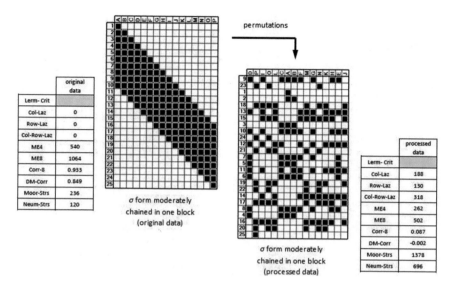

	original data
Lerm- Crit	
Col-Laz	0
Row-Laz	0
Col-Row-Laz	0
ME4	540
ME8	1064
Corr-8	0.933
DM-Corr	0.849
Moor-Strs	236
Neum-Strs	120

σ form moderately
chained in one block
(original data)

permutations

	processed data
Lerm- Crit	
Col-Laz	188
Row-Laz	130
Col-Row-Laz	318
ME4	262
ME8	502
Corr-8	0.087
DM-Corr	-0.002
Moor-Strs	1378
Neum-Strs	696

σ form moderately
chained in one block
(processed data)

Fig. 4.3 Data for one block moderately chained: $25x16 - 10$

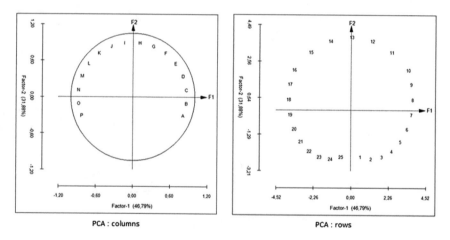

PCA : columns   PCA : rows

Fig. 4.4 PCA columns and rows respective representations

visually repairing an initial point of the circular geometric representation. Following
this choice, the circular sequence of points is scanned as a function of the polar
coordinates, or, more precisely, as a function of the polar angle.

The relevant scan can be performed in a clockwise or anticlockwise orientation,
depending on the distribution of points in the quadrant where the chosen initial point
is located. A possible correction modulo π can be required. Therefore, representations of such angular graphical paths are provided, associated with the PCA on both
columns and rows, bringing out their respective orders (see Fig. 4.6).

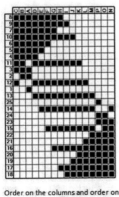

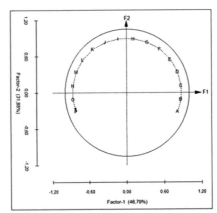

	PCA (F1)
Lerm- Crit	
Col-Laz	89
Row-Laz	6
Col-Row-Laz	95
ME4	436
ME8	768
Corr-8	0.574
DM-Corr	0.747
Moor-Strs	812
Neum-Strs	336

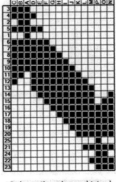

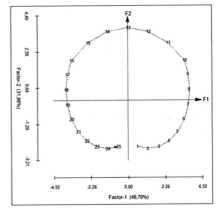

	PCA (F1-col)
Lerm- Crit	
Col-Laz	87
Row-Laz	6
Col-Row-Laz	93
ME4	432
ME8	768
Corr-8	0.572
DM-Corr	0.751
Moor-Strs	812
Neum-Strs	344

Order on the columns and order on the rows obtained by projection on the first factorial axis

Order on the columns obtained by projection on the first factorial axis and order on the rows driven by the order on the columns

Fig. 4.5 Seriation from the first PCA factor, followed by ROFCO

PCA (polar coordinates) : columns, with first column A

PCA (polar coordinates) : rows, with first row 1

Fig. 4.6 Polygonal lines following polar angular coordinates

These two orders on the columns and on the rows, coming from polar coordinates make it possible to recover the initial seriation.

This is also the case by considering the ROFCO algorithm applied to the order on the columns (see Sect. 5.2.3 of Chap. 5 and see Fig. 4.7).

Results when applying CA Method

In correspondence analysis, geometric representations of seriation shapes appear as parabolic graphs for both columns and rows. This is clearly shown in Fig. 4.8. As was the case in the PCA method, the respective projections on the first factorial axis, column and row elements determine a pair of respective orders on columns and rows. The crossing of these two orders gives a structure of seriation. Another

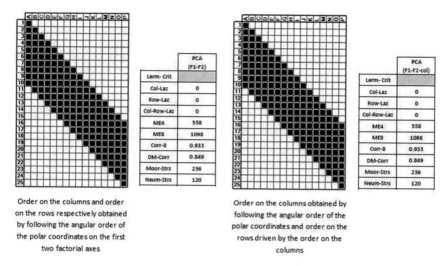

	PCA (F1-F2)
Lerm- Crit	
Col-Laz	0
Row-Laz	0
Col-Row-Laz	0
ME4	558
ME8	1098
Corr-8	0.933
DM-Corr	0.849
Moor-Strs	236
Neum-Strs	120

Order on the columns and order on the rows respectively obtained by following the angular order of the polar coordinates on the first two factorial axes

	PCA (F1-F2-col)
Lerm- Crit	
Col-Laz	0
Row-Laz	0
Col-Row-Laz	0
ME4	558
ME8	1098
Corr-8	0.933
DM-Corr	0.849
Moor-Strs	236
Neum-Strs	120

Order on the columns obtained by following the angular order of the polar coordinates and order on the rows driven by the order on the columns

Fig. 4.7 PCA Two techniques for recovering the seriation

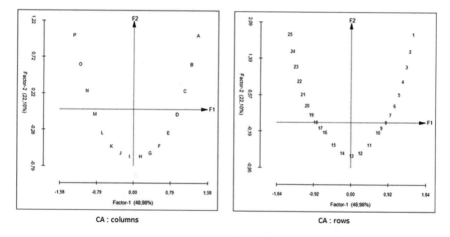

CA : columns

CA : rows

Fig. 4.8 CA Columns and rows respective representations

alternative is to keep only the column order and apply the ROFCO algorithm. These two techniques lead to the seriation from which we started (see Figs. 4.8 and 4.9).

As previously for the PCA method, a purely visual reading of the sequence of elements following the parabolic shape would allow us to retrieve the initial seriation. By using the technique of polar angular coordinates, we can automate this perception (see Fig. 4.10). On the other hand and similarly, we can only retain the order obtained on the columns and derive the corresponding order on the rows by using ROFCO. All of these different possibilities lead us to rediscover the initial form of the seriation (Fig. 4.11).

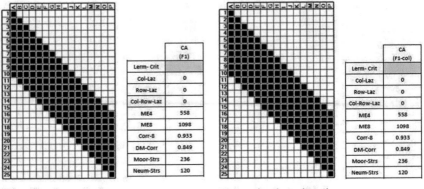

Order on the columns and order on the rows obtained by projection on the first factorial axis

Order on the columns obtained by projection on the first factorial axis and order on the rows driven by the order on the columns

Fig. 4.9 Seriation from the first CA factor, followed by ROFCO

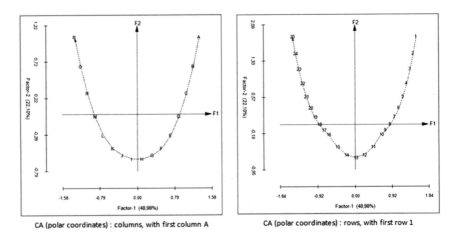

CA (polar coordinates) : columns, with first column A

CA (polar coordinates) : rows, with first row 1

Fig. 4.10 Polygonal lines following polar angular coordinates

Results when applying APS Method

In the geometric representation provided by this new method (see Sect. 2.4.5 in Chap. 2), the first two axes are driven by the first two poles of attraction. They are not orthogonal (see Sect. 2.4.5 of Chap. 2). We can nevertheless continue to use the principle of projection on the first axis parallelly to the second axis and thus derive an order on all the elements represented. However, the representations correspond to two independent treatments. There is no link between the pole of the first axis for the columns and that for the rows (Fig. 4.12).

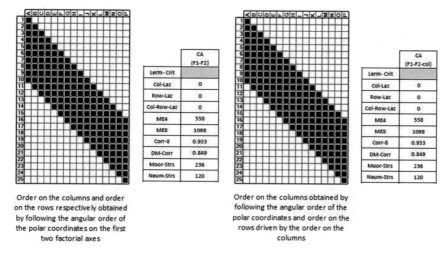

Order on the columns and order
on the rows respectively obtained
by following the angular order of
the polar coordinates on the first
two factorial axes

Order on the columns obtained by
following the angular order of the
polar coordinates and order on the
rows driven by the order on the
columns

Fig. 4.11   CA two techniques for recovering the seriation

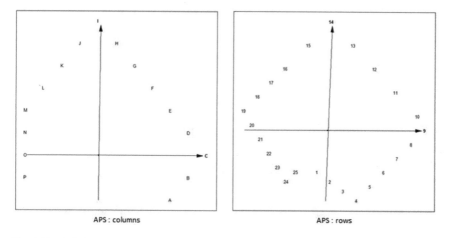

APS: columns   APS: rows

Fig. 4.12   APS Columns and rows respective representations

Here again, by relying on the only order of the columns, resulting from their projection on the first axis, we can observe the contribution of the ROFCO algorithm (see Sect. 5.2.3 in Chap. 5), to order the rows and thus, to recover the seriation (Fig. 4.13).

As for factorial analyzes, seriation can be brought out by visually following in an oriented direction the sequence of elements on the graphics of the geometrical representation. Also here the idea of automatically extracting an order from the angular values of the polar coordinates can be applied and this, on the column or row elements. Recall that the starting point for this, consists of choosing an origin element and a clockwise or counterclockwise course, these choices being visual.

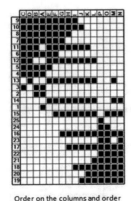

	APS (axe1)
Lerm- Crit	
Col-Laz	90
Row-Laz	12
Col-Row-Laz	102
ME4	420
ME8	732
Corr-8	0.523
DM-Corr	0.727
Moor-Strs	884
Neum-Strs	368

Order on the columns and order
on the rows obtained by
projection on the axis driven by
the first pole

	APS (axe1-col)
Lerm- Crit	
Col-Laz	22
Row-Laz	12
Col-Row-Laz	34
ME4	498
ME8	938
Corr-8	0.768
DM-Corr	0.804
Moor-Strs	520
Neum-Strs	228

Order on the columns obtained
by projection on the first pole axis
and order on the rows driven by
the order on the columns

Fig. 4.13 Seriation from the first attraction pole, followed by ROFCO

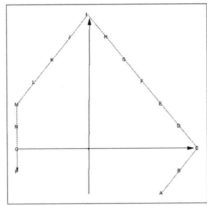

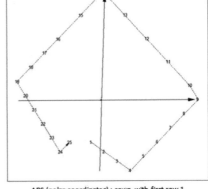

APS (polar coordinates) : columns, with first column A   APS (polar coordinates) : rows, with first row 1

Fig. 4.14 Polygonal lines following polar angular coordinates

In this framework, by using the only column order, we obtain—as in the case of factorial analyses—through ROFCO—the seriation.

By its nature, the Attraction Poles method based on similarities is very flexible. It only requires the definition of a similarity coefficient on the set to be processed (columns or rows of an incidence data table in our case) (Fig. 4.14).

In this context, we only have worked on the similarity indices expressed exhaustively in Sect. 6.2.1 of Chap. 6 and which resume the development of Sect. 2.3 of Chap. 2. More precisely, it is the coefficient (6.6) which is reduced to (6.9) for the case of comparing column attributes. On the other hand, the comparison of row objects uses (6.10) (Fig. 4.15).

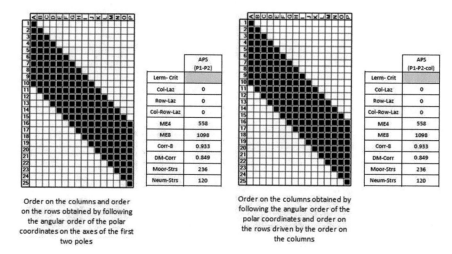

Order on the columns and order
on the rows obtained by following
the angular order of the polar
coordinates on the axes of the first
two poles

Order on the columns obtained by
following the angular order of the
polar coordinates and order on
the rows driven by the order on
the columns

Fig. 4.15 APS Two techniques for recovering the seriation

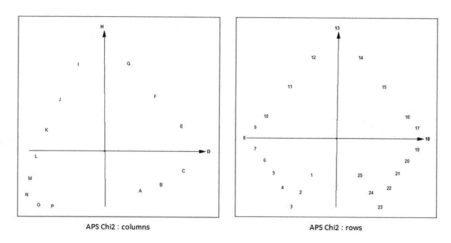

APS Chi2 : columns   APS Chi2 : rows

Fig. 4.16 APS-Chi2 columns and rows respective representations

An original index results from the adaptation of the correlation coefficient deriving from the Chi2 metric (4.9), the latter being developed in the correspondence analysis method (Figs. 4.16 and 4.17).

Results when applying MDS-HS Method

Multidimensional scaling brings out a seriation structure through a planar geometrical representation. Such a structure appears as a "horse-shoe" figure. In this method, the basic data is a distance matrix between the elements of the set **E** to be represented. In our concern, **E** is defined by the set of column attributes or that of

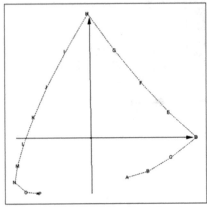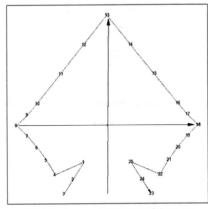

APS Chi2 (polar coordinates) : columns, with first column A   APS Chi2 (polar coordinates) : rows, with first row 3

Fig. 4.17 Polygonal lines following polar angular coordinates

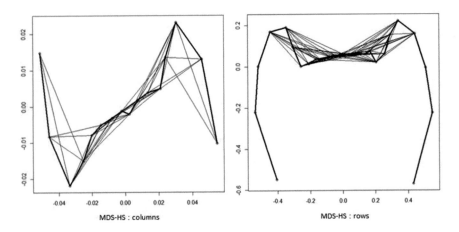

MDS-HS : columns   MDS-HS : rows

Fig. 4.18 MDS-HS columns and rows respective representations

row objects. In the case where a similarity index is defined on **E**, a distance index is associated with it in a coherent way (see Eqs. (6.6)–(6.12) in Chap. 6).

Let us emphasize that we have clearly distinguished the hamiltonian paths, enabling the respective orderings of columns and rows, associated with the seriation, to be obtained (Fig. 4.18).

By crossing these two orders (on the columns and on the rows), the seriation form of the data table is recovered (Fig. 4.19).

As is the case for the other methods, we obtain this last seriation structure from the only column order and via the ROFCO algorithm (Fig. 4.20).

Results following the use of OP2-S chaining algorithm

By examining the four respective representations of the column attributes given by the four methods (PCA, CA, APS and MDS-HS), we can observe that the start-

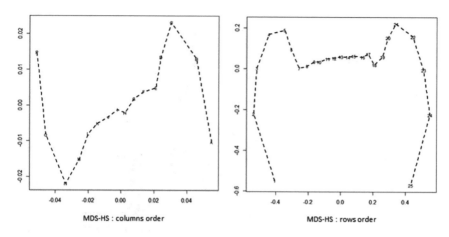

Fig. 4.19 MDS-HS hamiltonian paths

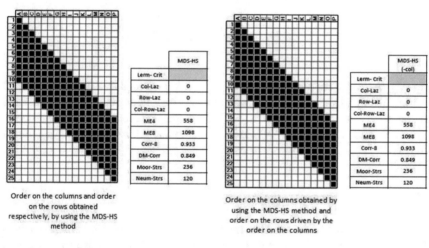

Order on the columns and order on the rows obtained respectively, by using the MDS-HS method

	MDS-HS
Lerm- Crit	
Col-Laz	0
Row-Laz	0
Col-Row-Laz	0
ME4	558
ME8	1098
Corr-8	0.933
DM-Corr	0.849
Moor-Strs	236
Neum-Strs	120

Order on the columns obtained by using the MDS-HS method and order on the rows driven by the order on the columns

	MDS-HS (-col)
Lerm- Crit	
Col-Laz	0
Row-Laz	0
Col-Row-Laz	0
ME4	558
ME8	1098
Corr-8	0.933
DM-Corr	0.849
Moor-Strs	236
Neum-Strs	120

Fig. 4.20 MDS-HS Seriations from two algorithmic alternatives

ing point of the polygonal line appears clearly to be A. In these conditions, the OP2-S chaining algorithm—depending on the origin element—gives a unique result regardless of the method employed and then, the obtained seriation through ROFCO algorithm is also unique. We consider here the result from the APS algorithm.

In the approach developed, only the columns geometric representation is exploited. This, because the specification of the OP2-S algorithm is relative to seriation on the column set. Finally, the result obtained is an ordered data table (columns and rows) according to its seriation structure, the order on the rows being derived from that on the columns through ROFCO algorithm (Fig. 4.21).

All the treatments carried out for the moderately chained sigma form were applied to all of the other σ forms. In the case of a single block, it is, on the one hand, the

Fig. 4.21 (PCA,CA,APS,MDS-HS) seriation from an extreme element of geometric representation of the columns

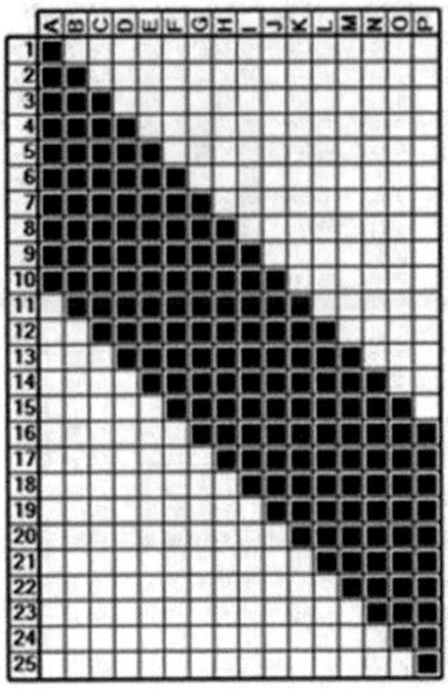

	APS (OP2-S)
Lerm-Crit	10.30
Col-Laz	0
Row-Laz	0
Col-Row-Laz	0
ME4	558
ME8	1098
Corr-8	0.933
DM-Corr	0.849
Moor-Strs	236
Neum-Strs	120

Order on the columns obtained by the OP2-S algorithm, starting with column A from APS representation and order on the rows driven by the order on the columns

strongly chained form and on the other, the weakly chained one (see Sects. 4.4.2.3 and 4.4.2.4). Also, the two-block σ form and the three-block σ form have been taken into account. Thus, we will be much shorter in the presentation of the results concerning these simulated sigma forms (see Sect. 4.4.2.5 and 4.4.2.6). The analysis of real cases is more specific and then will merit significant development (see Sect. 4.4.3).

4.4.2.3 σ Form Strongly Chained in One Block

As is the case for the sigma form in a moderately chained block, the strongly chained block can be interpreted in terms of a parallelogram. The respective parameters of the incidence data table are n = 21 for the row number, p = 11 for the column number and e = 21 for the number of one value along each of the columns. This last parameter can be interpreted, in a way, as a depth.

The original incidence data table is on the left of Fig. 4.22. That processed is directly on the right of the latter. It results from a pair of independent random permutations on the rows and on the columns of the initial table.

In accordance with what we have just indicated, we will limit ourselves here to showing only the planar geometric representations of all the column attributes, using the various methods. For a given method, this representation makes it possible to discover an order on the columns. In these conditions, the ROFCO algorithm (introduced in Sect. 4.3 and detailed in 5.2.3 of Chap. 5) enables the associated order on the rows to be discovered. In fact, it turned out that, for the data considered, all the methods used have the same result for the order on the columns.

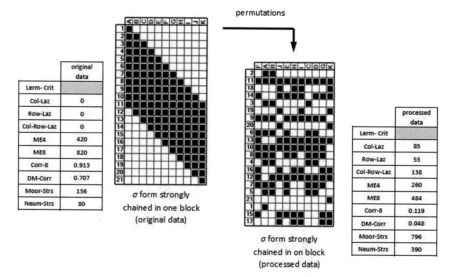

Fig. 4.22   The data processed and the associated criteria

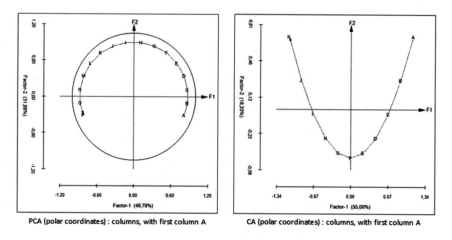

PCA (polar coordinates) : columns, with first column A   CA (polar coordinates) : columns, with first column A

Fig. 4.23   PCA and CA polygonal lines with polar angular scanning

Results when Applying PCA and CA Methods

In the geometric graphical representations given in the following, a polygonal line scaning the different points as a function of angular polar coordinates is employed. An initial point is detected visually (Fig. 4.23).

Results when Applying APS and MDS-HS Methods

As before, these results are obtained by following a polygonal line scaning the various points as a function of angular polar coordinates and starting with an original element, detected visually (Fig. 4.24).

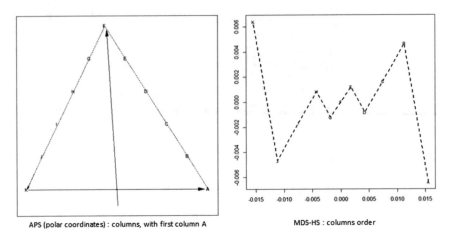

APS (polar coordinates) : columns, with first column A   MDS-HS : columns order

Fig. 4.24 Attraction poles APS and MDS-'horse-shoe' methods

	APS (OP2-S)
Lerm- Crit	7.10
Col-Laz	0
Row-Laz	0
Col-Row-Laz	0
ME4	420
ME8	820
Corr-8	0.913
DM-Corr	0.707
Moor-Strs	156
Neum-Strs	80

Order on the columns obtained by the OP2-S algorithm, starting with column A from APS representation and order on the rows driven by the order on the columns

Fig. 4.25 OP2-S algorithm starting with the origin point A from APS representation

Results Following the Use of OP2-S Chaining Algorithm

The four geometric representations of the columns provide the same source element of the concerned graph. That is to say A. Therefore, since the progress of OP2-S depends only on the initial point, all the graphical representations give a single result for the order on the columns. The one that is represented is associated with the APS representation (Fig. 4.25).

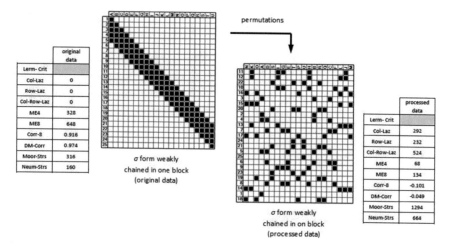

Lerm- Crit	original data
Col-Laz	0
Row-Laz	0
Col-Row-Laz	0
ME4	328
ME8	648
Corr-8	0.916
DM-Corr	0.974
Moor-Strs	316
Neum-Strs	160

σ form weakly
chained in one block
(original data)

permutations

Lerm- Crit	processed data
Col-Laz	292
Row-Laz	232
Col-Row-Laz	524
ME4	68
ME8	134
Corr-8	-0.101
DM-Corr	-0.049
Moor-Strs	1294
Neum-Strs	664

σ form weakly
chained in on block
(processed data)

Fig. 4.26 The data processed and the associated criteria

4.4.2.4 σ Form Weakly Chained in One Block

The ordered incidence data table according to its seriation structure is shown on the left. The parameters n, p and e are resp., equal to 25, 21 and 5. Methods techniques and algorithms are performed on a permuted version of this table. This version, shown on the right, is obtained from a pair of independent permutations on the rows and on the columns of the initial table. Each of both versions (the original and the permuted one) is accompanied by its local criteria.

As previously, only geometric planar representation of the columns is figured. A total order on the points—representing the columns—is deduced. This, according to following a rounded curve traversing all of the points. This curve—described from a starting point visually recognized—begins in a concave way and ends in a convex one (or, begins in a convex way and ends in a concave one). This order enables to infer an associated order on the rows, by the ROFCO algorithm (Fig. 4.26).

Results when Applying PCA and CA Methods

PCA, CA and MDS-HS give—through angular polar scanning—the same order on the columns. In the APS algorithm the twelve first elements from A to L are very clearly distinguished in the correct order. The nine other elements constitute an amalgamation which follows. Clearly, we can imagine to complete the whole of the seriation by carrying out the same algorithm on the subtable restricted to the columns from M to U and to the rows from 13 to 19. In fact, it should be remembered that the Theorems 2 and 3 concern only the strong chained sigma form. And, it is already remarkable that in the moderately chained case, the whole of the seriation could be restored (see Sect. 4.4.2.2) (Fig. 4.27).

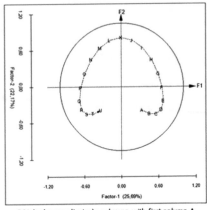

PCA (polar coordinates) : columns, with first column A

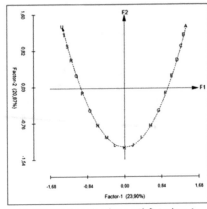

CA (polar coordinates) : columns, with first column A

Fig. 4.27 PCA and CA methods results

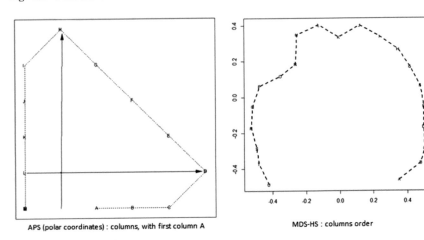

APS (polar coordinates) : columns, with first column A

MDS-HS : columns order

Fig. 4.28 APS and MDS-HS methods results

Results when Applying APS and MDS-HS Methods (Fig. 4.28)

Reconstruction

The following incidence table is obtained from the order on the column set provided by the geometric representation of the APS method. As usual, the resulting order on the rows is deduced from the application of ROFCO algorithm. Once again, let us insist on the fact that the simulated sigma form concerned, is too weakly chained for APS in order to be able to lead in a single step to the entire seriation. One specificity and therefore an originality of APS consists of requiring very little methodological input (Fig. 4.29).

Fig. 4.29 Reconstruction of
the seriation from APS

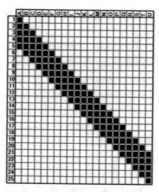

	APS (P1-P2-col)
Lerm- Crit	
Col-Laz	21
Row-Laz	40
Col-Row-Laz	61
ME4	270
ME8	532
Corr-8	0.782
DM-Corr	0.902
Moor-Strs	546
Neum-Strs	276

Order on the columns obtained by
following the angular order of the
polar coordinates and order on
the rows driven by the order on
the columns

Fig. 4.30 Applying OP-2S
algorithm from APS

	APS (OP2-S)
Lerm- Crit	11.15
Col-Laz	0
Row-Laz	0
Col-Row-Laz	0
ME4	328
ME8	648
Corr-8	0.916
DM-Corr	0.974
Moor-Strs	316
Neum-Strs	160

Order on the columns obtained by
the OP2-S algorithm, starting with
column A from APS representation
and order on the rows driven by the
order on the columns

Applying OP2-S algorithm

As already expressed several times (see its introduction in the Sect. 4.3 above
and its development in Sect. 5.3.1) of Chap. 5, the OP2-S algorithm needs a starting
point for its development. Relative to the order of the columns, the plane geometric
representation by a multidimensional data analysis method provides this last point
as the origin of this representation. Each of the four methods (PCA, CA, MDS-HS
and APS) gives A as a starting point. Under these conditions, it suffices to retain the
basic representation method APS to obtain Fig. 4.30.

4.4.2.5 σ Form into Two Blocks Strongly Chained

The simulated data table is on the left of Fig. 4.31. It comprises 42 rows and 22 columns. Each of the two blocks is strongly chained and can be circumscribed in an incidence data table with 21 rows and 11 columns. The first table includes the rows 1 to 21 and the columns A1 to K1. The second table includes the rows 22 to 42 and the columns A2 to K2.

As usual, the data table processed (on the right of Fig. 4.31) assumes practiced a pair of random permutations on rows and columns of the original data. A criteria table is attached with each version of the data table (Fig. 4.32).

As previously, only planar geometrical representation of column attributes is shown (see for example the preceding Sect. 4.4.2.4). The associated ordering on the row objects is then driven from ROFCO algorithm.

The methods developed so far relate to a single chained block. The latter corresponds to a related form on the set of column attributes. In other words, we assume the existence of an order on all the columns such that, relative to two consecutive columns for this order, the subset of the objects where the two attributes concerned are present, is not empty.

Here it is different. The simulated and seriated incidence table comprises more than a single block. In addition the different blocks are disjoint. In the case of two disjoint blocks, an object having an attribute corresponding to a column of one of the blocks does not have any attribute of the other block.

Nevertheless, the construction of a seriation is possible and that, by ordering the different blocks and by ordering the elements of each of the blocks, each one in relation to those which preceded it.

It may happen that the seriation can be directly obtained (see the case of the PCA method below). However, the most efficient approach is to cluster all of the columns in order to detect the most significant partition. This approach can be materialized by applying the LLA hierarchical clustering method and recognizing its most significant level (see Chaps. 9 and 10 in [6]).

In these conditions, a seriation will be carried out on each of the blocks and then, we rank the blocks. The number of blocks being relatively small, we can test all of the block orders, in such a way to retain the adequate one.

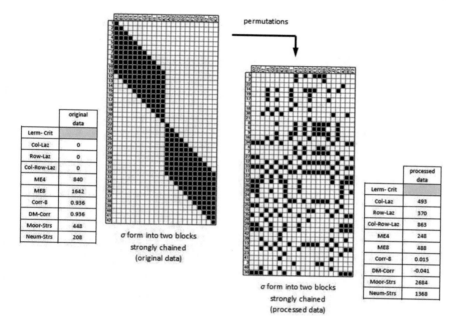

Fig. 4.31   The data processed and the associated criteria

Results when Applying Principal Component Analysis (PCA) Method

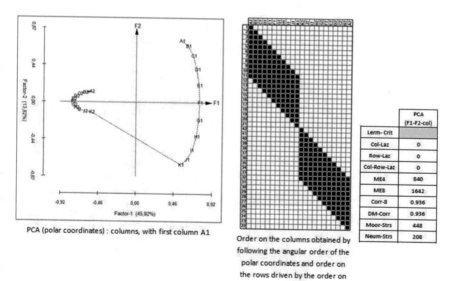

Fig. 4.32   Results of PCA method

Results when Applying Correspondence Analysis (CA) Method (Fig. 4.33)

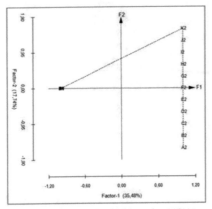

CA (polar coordinates) : columns, with first column A2

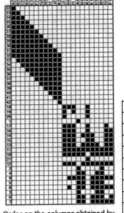

	CA (F1-F2-col)
Lerm- Crit	
Col-Laz	45
Row-Laz	46
Col-Row-Laz	91
ME4	726
ME8	1364
Corr-8	0.797
DM-Corr	0.889
Moor-Strs	968
Neum-Strs	424

Order on the columns obtained by following the angular order of the polar coordinates and order on the rows driven by the order on the columns

Fig. 4.33 Results of CA method

Results when Applying Attraction Poles Similarities (APS) Method (Fig. 4.34)

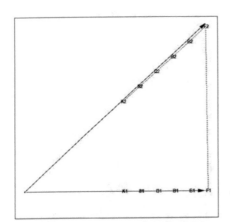

APS (polar coordinates) : columns, with first column A1

	APS (P1-P2-col)
Lerm- Crit	
Col-Laz	96
Row-Laz	90
Col-Row-Laz	186
ME4	526
ME8	1158
Corr-8	0.654
DM-Corr	0.830
Moor-Strs	1416
Neum-Strs	836

Order on the columns obtained by following the angular order of the polar coordinates and order on the rows driven by the order on the columns

Fig. 4.34 Results of APS method

Results when Applying Multidimensional Scaling 'Horse-shoe' Method
(Fig. 4.35)

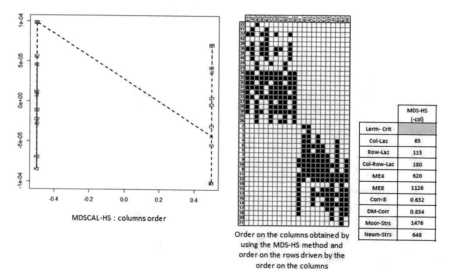

	MDS-HS (-col)
Lerm- Crit	
Col-Laz	65
Row-Laz	115
Col-Row-Laz	180
ME4	620
ME8	1126
Corr-8	0.632
DM-Corr	0.834
Moor-Strs	1476
Neum-Strs	648

MDSCAL-HS : columns order

Order on the columns obtained by
using the MDS-HS method and
order on the rows driven by the
order on the columns

Fig. 4.35 Results of MDS-HS method

Results Following the Use of OP2-S Chaining Algorithm
The respective elements initiating the preceding different geometric representations are not identical. Therefore, different options for applying OP2-S algorithm are
given in the following figures (Figs. 4.36 and 4.37).

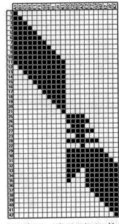

	PCA (OP2-S)
Lerm- Crit	12.17
Col-Laz	12
Row-Laz	6
Col-Row-Laz	18
ME4	810
ME8	1564
Corr-8	0.894
DM-Corr	0.928
Moor-Strs	604
Neum-Strs	268

Order on the columns obtained by the OP2-S algorithm, starting with column A1 deduced from PCA representation and order on the rows driven by the order on the columns

	CA (OP2-S)
Lerm- Crit	12.38
Col-Laz	2
Row-Laz	1
Col-Row-Laz	3
ME4	834
ME8	1628
Corr-8	0.928
DM-Corr	0.935
Moor-Strs	476
Neum-Strs	220

Order on the columns obtained by the OP2-S algorithm, starting with column A2 deduced from CA representation and order on the rows driven by the order on the columns

Fig. 4.36 Results of OP2-S from PCA and CA methods

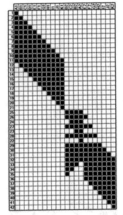

	APS (OP2-S)
Lerm- Crit	12.17
Col-Laz	12
Row-Laz	6
Col-Row-Laz	18
ME4	810
ME8	1564
Corr-8	0.894
DM-Corr	0.928
Moor-Strs	604
Neum-Strs	268

Order on the columns obtained by the OP2-S algorithm starting with column A1 from APS representation and order on the rows driven by the order on the columns

	MDS-HS (OP2-S)
Lerm- Crit	11.76
Col-Laz	27
Row-Laz	16
Col-Row-Laz	43
ME4	774
ME8	1466
Corr-8	0.838
DM-Corr	0.913
Moor-Strs	770
Neum-Strs	330

Column order obtained by the OP2-S algorithm starting with column F2 from MDS-HS representation and row order driven by the order on the columns

Fig. 4.37 Results of OP2-S from APS and MDS-HS methods

4.4.2.6 σ Form into Three Blocks Strongly Chained

The simulated data table is on the left of Fig. 4.38. It comprises 45 rows and 24 columns. Each of the three blocks is strongly chained and can be circumscribed in an incidence data table with 15 rows and 8 columns. The first table includes the rows 1 to 15 and the columns A1 to A8, the second table the rows 16 to 30 and the columns B1 to B8, the third one, the rows 31 to 45 and the columns, C1 to C8.

As previously, only planar geometrical representation of the columns is given. The latter provides an order on the column set. Ordering the rows is driven from applying ROFCO algorithm.

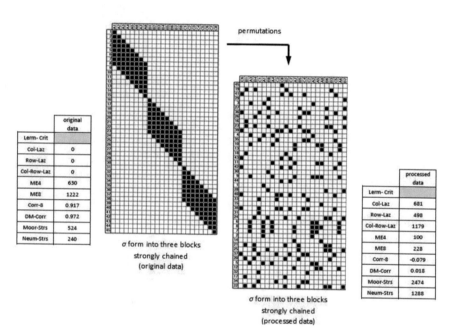

	original data
Lerm- Crit	
Col-Laz	0
Row-Laz	0
Col-Row-Laz	0
ME4	630
ME8	1222
Corr-8	0.917
DM-Corr	0.972
Moor-Strs	524
Neum-Strs	240

σ form into three blocks strongly chained (original data)

permutations

	processed data
Lerm- Crit	
Col-Laz	681
Row-Laz	498
Col-Row-Laz	1179
ME4	100
ME8	228
Corr-8	-0.079
DM-Corr	0.018
Moor-Strs	2474
Neum-Strs	1288

σ form into three blocks strongly chained (processed data)

Fig. 4.38 The data processed and the associated criteria

Applying Principal Component Analysis (PCA) (Fig. 4.39)

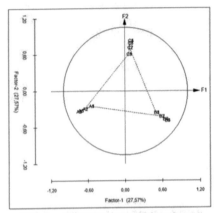

PCA (polar coordinates) : columns, with first column B4

Order on the columns obtained by
following the angular order of the
polar coordinates and order on the
rows driven by the order on the
columns

	PCA (F1-F2-col)
Lerm- Crit	
Col-Laz	184
Row-Laz	292
Col-Row-Laz	476
ME4	382
ME8	738
Corr-8	0.581
DM-Corr	0.628
Moor-Strs	1488
Neum-Strs	736

Fig. 4.39 Results of PCA method

Applying Correspondence Analysis (CA) (Fig. 4.40)

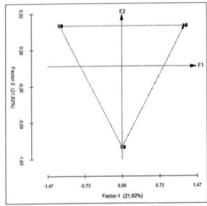

CA (polar coordinates) : columns, with first column A1

Order on the columns obtained by
following the angular order of the
polar coordinates and order on the
rows driven by the order on the
columns

	CA (F1-F2-col)
Lerm- Crit	
Col-Laz	89
Row-Laz	203
Col-Row-Laz	292
ME4	422
ME8	758
Corr-8	0.593
DM-Corr	0.772
Moor-Strs	1426
Neum-Strs	648

Fig. 4.40 Results of CA method

Applying Attraction Poles Similarities method (APS) (Fig. 4.41)

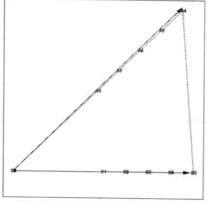

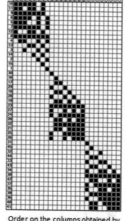

APS (polar coordinates) : columns, with first column B1

	APS (P1-P2-col)
Lerm- Crit	
Col-Laz	74
Row-Laz	62
Col-Row-Laz	136
ME4	420
ME8	826
Corr-8	0.640
DM-Corr	0.930
Moor-Strs	1292
Neum-Strs	652

Order on the columns obtained by following the angular order of the polar coordinates and order on the rows driven by the order on the columns

Fig. 4.41   Results of APS method

Applying   Multidimensional   Scaling   'horse-shoe'   method   (MDS-HS) (Fig. 4.42)

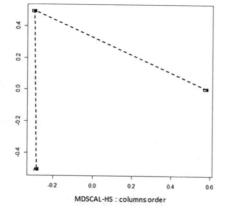

MDSCAL-HS : columns order

	MDS-HS (-col)
Lerm- Crit	
Col-Laz	69
Row-Laz	63
Col-Row-Laz	132
ME4	416
ME8	866
Corr-8	0.677
DM-Corr	0.942
Moor-Strs	1182
Neum-Strs	650

Order on the columns obtained by using the MDS-HS method and order on the rows driven by the order on the columns

Fig. 4.42   Results of MDS-HS method

Applying OP2-S Ordering Algorithm

This algorithm requires an origin element. Relative to a planar geometrical representation of the column attributes by multidimensional data analysis, this element is detected at an extremity of the representation. This element is not necessarily the same for each of the methods used and then, we consider different options (Figs. 4.43 and 4.44).

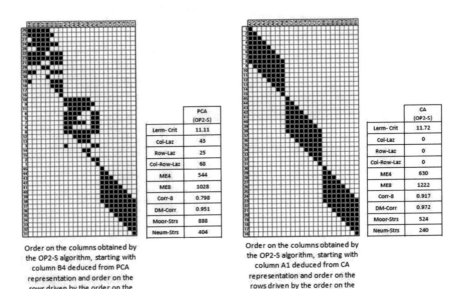

	PCA (OP2-S)
Lerm- Crit	11.11
Col-Laz	43
Row-Laz	25
Col-Row-Laz	68
ME4	544
ME8	1028
Corr-8	0.798
DM-Corr	0.951
Moor-Strs	888
Neum-Strs	404

	CA (OP2-S)
Lerm- Crit	11.72
Col-Laz	0
Row-Laz	0
Col-Row-Laz	0
ME4	630
ME8	1222
Corr-8	0.917
DM-Corr	0.972
Moor-Strs	524
Neum-Strs	240

Order on the columns obtained by the OP2-S algorithm, starting with column B4 deduced from PCA representation and order on the rows driven by the order on the columns

Order on the columns obtained by the OP2-S algorithm, starting with column A1 deduced from CA representation and order on the rows driven by the order on the columns

Fig. 4.43 Results of OP2-S from PCA and CA methods

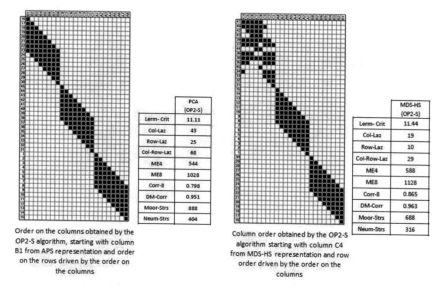

	PCA (OP2-S)
Lerm- Crit	11.11
Col-Laz	43
Row-Laz	25
Col-Row-Laz	68
ME4	544
ME8	1028
Corr-8	0.798
DM-Corr	0.951
Moor-Strs	888
Neum-Strs	404

Order on the columns obtained by the OP2-S algorithm, starting with column B1 from APS representation and order on the rows driven by the order on the columns

	MDS-HS (OP2-S)
Lerm- Crit	11.44
Col-Laz	19
Row-Laz	10
Col-Row-Laz	29
ME4	588
ME8	1128
Corr-8	0.865
DM-Corr	0.963
Moor-Strs	688
Neum-Strs	316

Column order obtained by the OP2-S algorithm starting with column C4 from MDS-HS representation and row order driven by the order on the columns

Fig. 4.44 Results of OP2-S from APS and MDS-HS methods

4.4.3 Real Data

Two real and famous archaeological data sets are considered. The first one concerns the analysis of Merovingian belt buckles-plates and the second, a study on the Münsingen celtic necropolis.

The respective evaluations of the automated seriation methods can be, to a certain extent, related to the values of different criteria (see Sect. 3.3.4 of Chap. 3 and Sect. 4.3). However, and ultimately, these evaluations are closely linked to the researcher's knowledge, the latter being influenced by certain preconceived ideas linked to validation of hypotheses. Let us here return to the final part of Sect. 4.3 where we recalled the correction made by an archaeologist (Hodson) relative to the result provided by a mathematician (Kendall). And in fact, all the results we get by processing real data are left for expert review.

4.4.3.1 Merovingian Bukles-Plates

The dataset is about decoration types of belt buckles-plates from the Merovingian periods, from the end of the 6th to the beginning of the 8th century, found in tombs in northeastern France. The data includes 59 buckles-plates and 26 decor types giving rise to a datatable comprising 59 rows and 26 columns(see Fig. 4.46).

An example of a belt buckle-plate decorated with silver damasquinure is given in Fig. 4.45.

Example of a 7th century decorated belt buckle-plate.

Fig. 4.45 A belt buckle-plate

First, the data was processed manually by Patrick Perin [10, 11], researcher in the Bertin's laboratoty (see Sect. 3.2.1 of Chap. 3) (see Fig. 4.46). In a manual approach, two trends can implicitly intervene for the mutual organization of the rows and columns of the data table. The first is grouping and therefore, refers to classification, the second is ordinal and therefore, refers to seriation [5, 12] Chap. 18, [13].

In the algorithms developed in this work, it is the ordinal aspect which largely dominates. We focus on the search for a seriation translating in the case of our data here, a chronological evolution of the use of the decors. It is this last point that we take up and illustrate. However, as seen in Sects. 4.4.2.5 and 4.4.2.6, in the case where a classification (clustering) structure underlies the data table, the latter appears.

As we have already expressed, it does not make sense to evaluate in a relative way the different seriation methods on the only basis of the values of the criteria tables. Certainly, these tables give a general indication. But, each method has its own logic which the archaeologist must take into account when interpreting the results.

It can be seen that the tables of criteria associated to the correspondence analysis (CA) and the multidimensional scaling 'horse-shoe' (MDS-HS) methods are very similar. These methods are paticularly developed and optimized (see Figs. 4.50 and 4.52). Note that correspondence analysis is popular with a certain audience for its ability to highlight evolutionary phenomena such as seriation [14]. Now, to compare mutually different serial orders, we can refer to the tools established in Chap. 6 of [6].

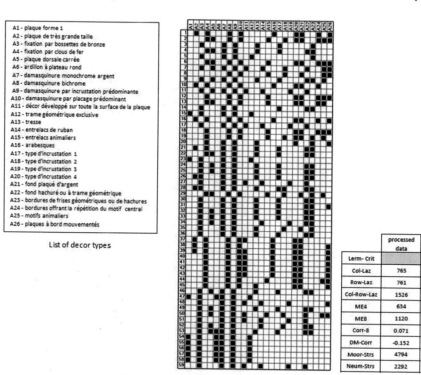

A1 - plaque forme 1
A2 - plaque de très grande taille
A3 - fixation par bossettes de bronze
A4 - fixation par clous de fer
A5 - plaque dorsale carrée
A6 - ardillon à plateau rond
A7 - damasquinure monochrome argent
A8 - damasquinure bichrome
A9 - damasquinure par incrustation prédominante
A10 - damasquinure par placage prédominant
A11 - décor développé sur toute la surface de la plaque
A12 - trame géométrique exclusive
A13 - tresse
A14 - entrelacs de ruban
A15 - entrelacs animaliers
A16 - arabesques
A17 - type d'incrustation 1
A18 - type d'incrustation 2
A19 - type d'incrustation 3
A20 - type d'incrustation 4
A21 - fond plaqué d'argent
A22 - fond hachuré ou à trame géométrique
A23 - bordures de frises géométriques ou de hachures
A24 - bordures offrant la répétition du motif central
A25 - motifs animaliers
A26 - plaques à bord mouvementés

List of decor types

	processed data
Lerm- Crit	
Col-Laz	765
Row-Laz	761
Col-Row-Laz	1526
ME4	634
ME8	1120
Corr-8	0.071
DM-Corr	-0.152
Moor-Strs	4794
Neum-Strs	2292

Merovingian decorated buckles-plates
(original data)

Fig. 4.46 The data processed

Relative to the various methods of column geometric representation (PCA, CA, APS, MDS-HS), consider the retention of the initial element which engages the representation via the polar angular scan. The application of the OP2-S algorithm (see 4.3 and more explicitly 5.3.1) from each of the four elements retained—relating resp., to the four methods—leads to very similar results in terms of the values of the criteria.

Fig. 4.47 Bertin's manual
treatment

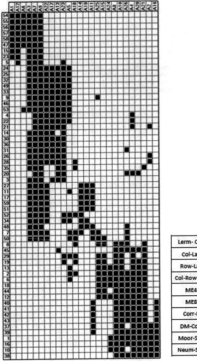

Manual processing (Bertin's laboratory)

	Bertin (manual)
Lerm- Crit	
Col-Laz	206
Row-Laz	228
Col-Row-Laz	434
ME4	1444
ME8	2702
Corr-8	0.816
DM-Corr	0.800
Moor-Strs	1704
Neum-Strs	696

Applying Principal Component Analysis (PCA) Method

$\overrightarrow{OA1}$ is the initial vector chosen to describe the sequence of points as a function of the polar angle. We could also have started from $\overrightarrow{OA8}$ where $A8$ is the element furthest to the right on the first factorial axis.

Applying Correspondence Analysis (CA) Method

In correspondence analysis, the order indicated by the first factor is close to that obtained by manual processing (see Figs. 4.47 and 4.49). Simply, the overall orientation is opposite in terms of rows and columns. This order appears very competitive with regard to all of the criteria (Fig. 4.48).

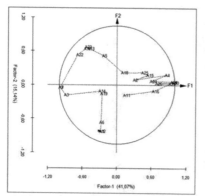

PCA (polar coordinates) : columns, with first column A11

PCA	
Lerm- Crit	(F1+F2-col)
Col-Laz	201
Row-Laz	233
Col-Row-Laz	434
ME4	1384
ME8	2598
Corr-8	0.778
DM-Corr	0.800
Moor-Strs	1888
Neum-Strs	808

Order on the columns obtained by following the angular order of the polar coordinates and order on the rows driven by the order on the columns

Fig. 4.48 Principal component analysis on merovingian data

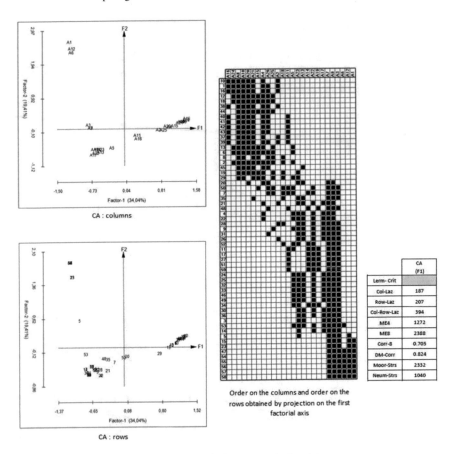

	CA (F1)
Lerm- Crit	
Col-Laz	187
Row-Laz	207
Col-Row-Laz	394
ME4	1272
ME8	2388
Corr-8	0.705
DM-Corr	0.824
Moor-Strs	2332
Neum-Strs	1040

CA : columns

CA : rows

Order on the columns and order on the rows obtained by projection on the first factorial axis

Fig. 4.49 First factor of correspondence analysis on merovingian data

In this version of the correspondence analysis application, we follow the points of the representation of the columns according to a scan, ordered by the polar angle and we restore the order on the rows from the ROFCO algorithm. The results are very similar to the preceding treatment (Fig. 4.50).

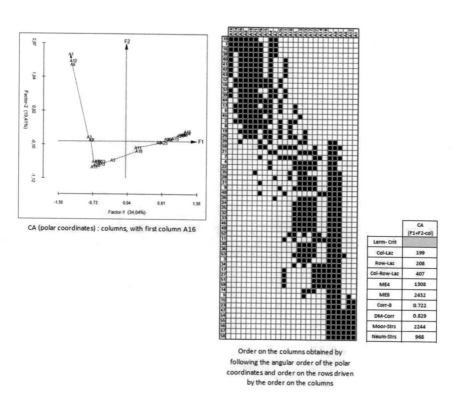

CA (polar coordinates) : columns, with first column A16

	CA (F1+F2-col)
Lerm- Crit	
Col-Laz	199
Row-Laz	208
Col-Row-Laz	407
ME4	1308
ME8	2432
Corr-8	0.722
DM-Corr	0.829
Moor-Strs	2244
Neum-Strs	968

Order on the columns obtained by
following the angular order of the polar
coordinates and order on the rows driven
by the order on the columns

Fig. 4.50 CA planar geometrical representation followed by ROFCO

Applying Attraction Poles Similarities (APS) Method

This method also brings out the seriation but with somewhat less optimized values of the criteria than in the case of correspondence analysis. Its value remains linked to the simplicity of the tool (Fig. 4.51).

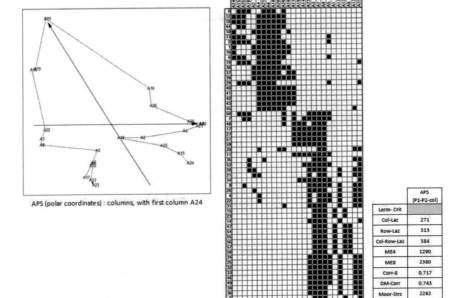

APS (polar coordinates) : columns, with first column A24

	APS (P1-P2-col)
Lerm- Crit	
Col-Laz	271
Row-Laz	313
Col-Row-Laz	584
ME4	1290
ME8	2380
Corr-8	0.717
DM-Corr	0.743
Moor-Strs	2262
Neum-Strs	976

Order on the columns obtained by
following the angular order of the polar
coordinates and order on the rows driven
by the order on the columns

Fig. 4.51 APS planar geometrical representation followed by ROFCO

Applying Multidimensional 'Horse-shoe' (MDS-HS) Method (Fig. 4.52)

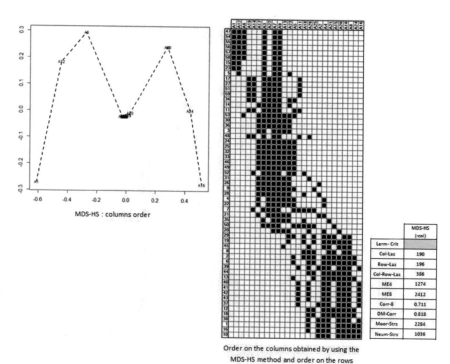

MDS-HS : columns order

Order on the columns obtained by using the
MDS-HS method and order on the rows
driven by the order on the columns

	MDS-HS (-col)
Lerm- Crit	
Col-Laz	190
Row-Laz	196
Col-Row-Laz	386
ME4	1274
ME8	2412
Corr-8	0.711
DM-Corr	0.818
Moor-Strs	2284
Neum-Strs	1036

Fig. 4.52 MDS-HS planar geometrical representation followed by ROFCO

Applying OP2-S Ordering Algorithm (Figs. 4.53, 4.54, 4.55 and 4.56)

Fig. 4.53 OP2-S from origin element in PCA representation

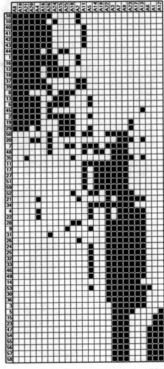

	PCA (OP2-S)
Lerm- Crit	12.78
Col-Laz	186
Row-Laz	251
Col-Row-Laz	437
ME4	1356
ME8	2518
Corr-8	0.755
DM-Corr	0.614
Moor-Strs	2006
Neum-Strs	850

Order on the columns obtained by the OP2-S algorithm, starting with column A8 deduced from PCA representation and order on the rows driven by the order on the columns

Fig. 4.54 OP2-S from origin element in CA representation

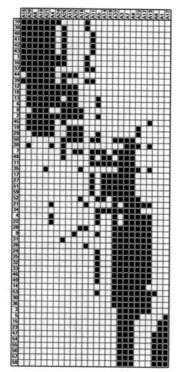

	CA (OP2-S)
Lerm- Crit	12.83
Col-Laz	176
Row-Laz	232
Col-Row-Laz	408
ME4	1348
ME8	2506
Corr-8	0.753
DM-Corr	0.821
Moor-Strs	2026
Neum-Strs	854

Order on the columns obtained by the OP2-S algorithm, starting with column A16 deduced from CA representation and order on the rows driven by the order on the columns

Fig. 4.55 OP2-S from origin element in APS representation

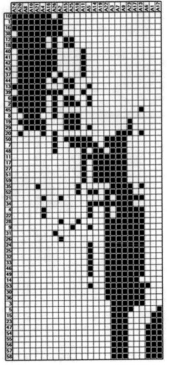

	APS (OP2-S)
Lerm- Crit	12.66
Col-Laz	187
Row-Laz	225
Col-Row-Laz	412
ME4	1358
ME8	2544
Corr-8	0.765
DM-Corr	0.820
Moor-Strs	1970
Neum-Strs	842

Order on the columns obtained by the OP2-S
algorithm, starting with column A24 deduced
from APS representation and order on the
rows driven by the order on the columns

Fig. 4.56 OP2-S from
origin element in MDS-HS
representation

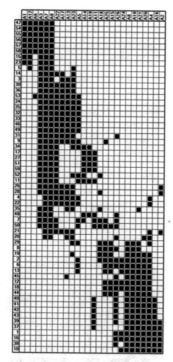

	MDS-HS (OP2-S)
Lerm- Crit	13.20
Col-Laz	176
Row-Laz	215
Col-Row-Laz	391
ME4	1422
ME8	2680
Corr-8	0.814
DM-Corr	0.826
Moor-Strs	1706
Neum-Strs	718

Order on the columns obtained by the OP2-S
algorithm, starting with column A1 deduced
from MDS-HS representation and order on the
rows driven by the order on the columns

4.4.3.2 The La Tène Cemetery at Münsingen-Rain

The study of this celtic cemetery of Münsingen in Switzerland gave rise to several experiments, discussions and publications. The data is very present in the R software. We situate ourselves directly in relation to the references [8, 15, 16].

Recall that this is the description of 59 graves by 70 Boolean attributes of presence absence. Each attribute is associated with the presence of an object in a grave. Essentially, these objects are defined by fibulae.

Before presenting results, let us note that many methodological remarks introducing the previous Sect. 4.4.3.1 are valid here.

We consider in Fig. 4.57 data in the R package. This is a result associated with Kendall's 'horse-shoe' method. The precise way to obtain it is not really explicit. It is surprising in itself. Nothing is said about the way of obtaining the column order. Indeed, in a large part above the diagonal of the reorganized data table, there is no presence value. For all methods practiced—including that of the 'horse-shoe'—this property does not occur. In fact, the possible occurrence of presence values above the diagonal is more natural and expected by the archaeologist compared to chronological evolution and succession. Let us emphasize here once again, that in our approach, determining the column order is the first and fundamental task. And in fact, for some options, row order can be driven.

Indeed, by considering the order of the columns in Fig. 4.57 and by applying the ROFCO algorithm, to associate the corresponding row order with it, perfect stability with the result provided from R, is not verified (see Fig. 4.58).

For reasons of readability and space, we will sometimes, simply show in what follows the incidence tables reorganized in columns and rows according to a seriation structure. Each of the tables is associated with a planar geometric representation by a data analysis method. The method used and its role will be explained. And indeed, the design followed in Sect. 4.4.3.1 is taken into account here.

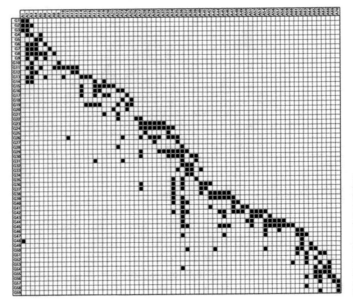

	munsingen. data (R)
Lerm- Crit	
Col-Laz	245
Row-Laz	402
Col-Row-Laz	647
ME4	478
ME8	866
Corr-8	0.567
DM-Corr	0.942
Moor-Strs	2574
Neum-Strs	1286

Fig. 4.57 Münsingen-Rain data as expressed in R

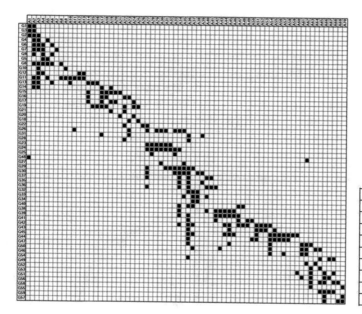

	munsingen. data (R) col
Lerm- Crit	
Col-Laz	251
Row-Laz	402
Col-Row-Laz	653
ME4	562
ME8	820
Corr-8	0.542
DM-Corr	0.951
Moor-Strs	2660
Neum-Strs	1236

Fig. 4.58 R data; order on the rows driven by ROFCO algorithm

Applying Principal Component Analysis (PCA) Method

As reported in Sect. 4.4.3.1, the PCA method is applied in order to provide a planar geometric representation of all column attributes. In the case of this data, the matrix of incidence—of presence-absence—includes an excessively low number of presence values. In addition, a large dispersion is observed for the number of presences and this, in the columns as in the rows. In this situation the PCA method turns out to be vulnerable. The graphic representation as it appears in Fig. 4.59 seems at first sight difficult to exploit. For this representation, a polygonal line starting with A2 and crossing all the points was indeed constructed according to a polar angular scan. This polygonal line enables an order on the columns to be determined. From the latter, an order on the rows can be deduced. In these conditions, the incidence table reordered in columns and rows reveals a seriation which has its own archaeological interest. However, relative to the ordered incidence table, the density of presence values along the diagonal of the reordered incidence data table, has appeared lower than in other methods such as correspondence analysis or multidimensional scaling.

In these conditions, we only kept the origin A2 of the polygonal line as it appears visually in Fig. 4.59 and this, to attach to it the OP2-S chaining algorithm (see the last paragraph of Sect. 4.3 for the mention of this algorithm). By this way, efficient result is obtained (see in the following of this section).

Now, as already mentioned in Sect. 4.2.1, correspondence analysis is aimed at the symmetric analysis of rows through columns and of columns through rows of a contingency table. This method is in fact a very specific and very original adaptation of principal component analysis [3, 4, 17, 18] Chap. 6 of [5, 14].

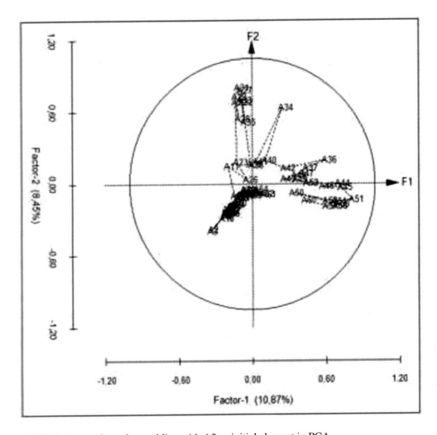

Fig. 4.59 Polar angular polygonal line with A2 as initial element in PCA

Applying Correspondence Analysis (CA) method

Let $\mathcal{N}(\mathbb{J})$ (resp. $\mathcal{N}(\mathbb{I})$) be the point cloud associated with the set of column attributes in a geometric space, as defined in the correspondence analysis (see Sects. 5.3.2.1 and 7.2.8.1 in [6]). Three versions of the latter method are applied. In each of them, the analysis of the cloud $\mathcal{N}(\mathbb{J})$ is used.

For the first version, we consider the respective projections on the first factorial axis of the cloud $\mathcal{N}(\mathbb{J})$ on the one hand and of that $\mathcal{N}(\mathbb{I})$ on the other hand (see Fig. 4.60). As a result, two orders on the columns and rows, resp., are obtained. Crossing these two orders leads to the seriation depicted in Fig. 4.61. The concentration of the presence values around the diagonal as well as the table of the values of the criteria show the quality of the seriation.

In the second version, matter is also a pair of plane factorial representations of $\mathcal{N}(\mathbb{J})$ on the one hand and of $\mathcal{N}(\mathbb{I})$ on the other. But here, the order on the columns as well as the order on the rows are obtained by means of a polar angular scan (see Fig. 4.62). It is of importance to notice that A1 (rep. G1) is chosen as the first element for the column representation (resp. row representation). The result is expressed in Fig. 4.63.

In the third version, only the plane factorial representation of all the columns is considered. This representation is polar angularly scanned. From the resulting order, the ROFCO algorithm proposes an order on the rows. Crossing the two orders leads to the reordered table which clearly reveals a seriation (see Fig. 4.64).

The three versions described are well optimized. This, relatively, on the one hand, to the values of the criteria and—in conjunction—to the diagonal concentration of the presence values in the reordered incidence table. We leave it to the reader to examine the mutual differences between the three results.

Fig. 4.60 Column and row
factorial geometric
representation in CA

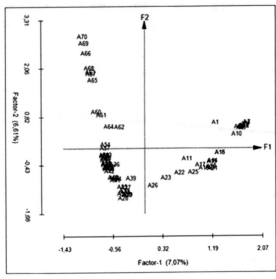

CA : columns

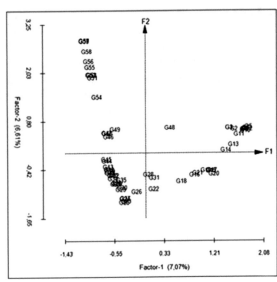

CA : rows

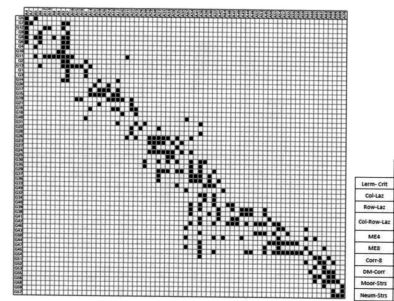

	CA (F1)
Lerm- Crit	
Col-Laz	260
Row-Laz	403
Col-Row-Laz	663
ME4	376
ME8	674
Corr-8	0.440
DM-Corr	0.959
Moor-Strs	2940
Neum-Strs	1404

Order on the columns and order on the rows obtained by projection on the first factorial axis

Fig. 4.61 Orderings columns and rows from the first factor in CA

Fig. 4.62 Angular scan for column and row factorial geometric representation in CA

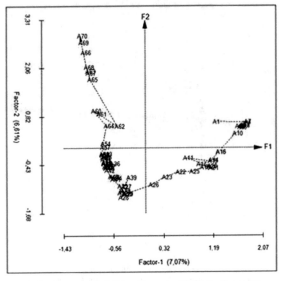

CA (polar coordinates) : columns, with first column A1

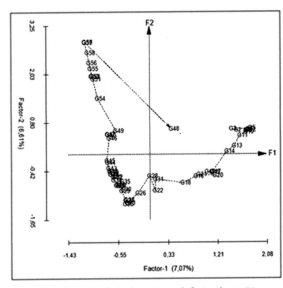

CA (polar coordinates) : rows. with first column G1

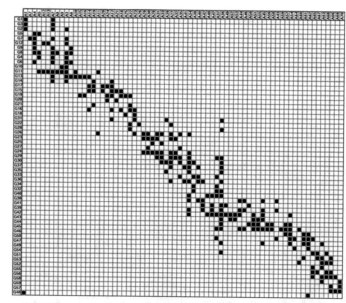

Order on the columns and order on the rows obtained by following the angular order of the polar coordinates

Fig. 4.63   Orderings columns and rows in CA

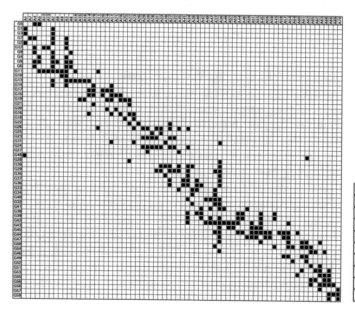

Fig. 4.64   Angular scan for ordering the columns followed by ROFCO in CA

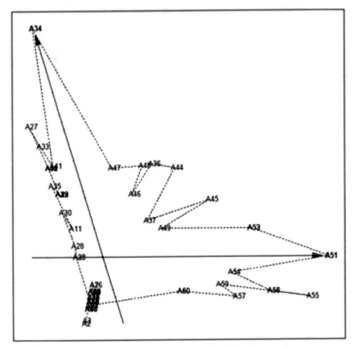

APS (polar coordinates) : columns, with first column A2

Fig. 4.65 Angular polygonal line with A2 as origin in APS

Applying Attraction Poles based on Similarities method

We emphasized in Chap. 2 on the methodological proximity between PCA and APS (see Sect. 2.4.6). The planar geometric representation of the columns is provided in Fig. 4.65. As for PCA, it is the point A2 which is considered as the point of origin from which the polygonal line describes all the points by a polar angular scan. As for PCA, we consider the sequence algorithm OP2-S from A2 to find the seriation.

Multidimensional Scaling 'Horse-shoe' method

The graphs obtained in the case of the application of the MDS-HS method: planar representations, each followed by scanning by a polygonal line crossing all the points, are given for the columns, as for the lines in Fig. 4.66. These lines allow the reordering of the columns and rows of the incidence data table. Both orderings are considered in Fig. 4.67. Column ordering followed by a driven row ROFCO ordering is used in Fig. 4.68. This reorganization reveals a seriation which appears to be of different quality than that which can be provided by the correspondence analysis. Nevertheless, for both cases the origin point of the column representation turns out to be A1.

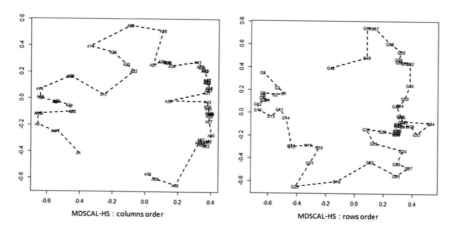

Fig. 4.66   Angular polygonal lines for row and column orderings in MDS-HS

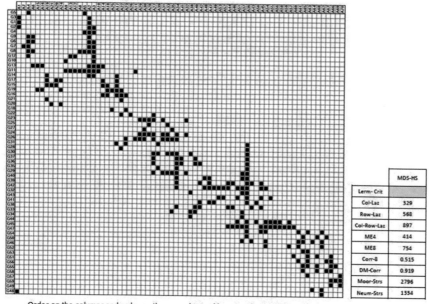

Order on the columns and order on the rows obtained by using the MDS-HS method

Fig. 4.67   Direct orderings rows and columns in MDS-HS

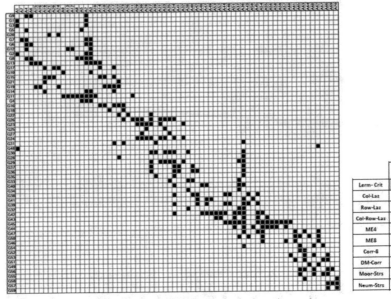

	MDS-HS (-col)
Lerm- Crit	
Col-Laz	339
Row-Laz	568
Col-Row-Laz	907
ME4	372
ME8	670
Corr-8	0.452
DM-Corr	0.940
Moor-Strs	2958
Neum-Strs	1416

Order on the columns obtained by using the MDS-HS method and order on the rows driven by the order on the columns

Fig. 4.68 Column order and driven row order by ROFCO in MDS-HS

Applying OP2-S Chaining Algorithm

We give here the results of the OP 2-S chaining algorithm for an initialization which is specified in a planar geometric representation of the columns. This is A2 for the ACP and APS methods (Fig. 4.69). This is A1 for the CA and MDS-HS methods (Fig. 4.70). Once the column order is obtained, the row order is deduced from ROFCO.

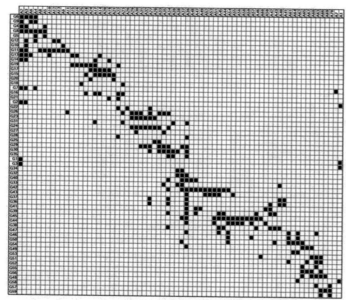

	PCA (OP2-S)
Lerm- Crit	20.12
Col-Laz	314
Row-Laz	649
Col-Row-Laz	963
ME4	438
ME8	754
Corr-8	0.509
DM-Corr	0.902
Moor-Strs	2736
Neum-Strs	1266

Order on the columns obtained by the OP2-S algorithm, starting with column A2 deduced from PCA and APS representations and order on the rows driven by the order on the columns

Fig. 4.69 OP2-S(A2) from common origin of PCA and APS graphical representations

4.4.4 Some Concluding Remarks

By considering all the results of the geometric representations of columns as a whole, it appears that CA and MDS-HS methods make it possible to bring out seriations. And this, both for the simulated data and real data.

In the case of simulated data, it is difficult to distinguish between the respective abilities of the different methods. For a single block all the methods have a correct behavior in the moderately chained and strongly chained versions.

In the case of more than one block, all the methods and particularly APS (see Figs. 4.34 and 4.41), prove to be able to recognize the existence and the number of blocks.

We can notice the interesting result of PCA method depicted in Fig. 4.32. However, as already mentioned, for weakly dense (few values equal to one) and scattered data as to the number of values equal to one per column, the PCA method can behave less optimally, compared to the correspondence analysis. This finding was observed in the case of Münsingen-Rain data. The extraction of the least loaded rows in terms of presence values leads to a significant improvement of the results. Research should continue in this regard.

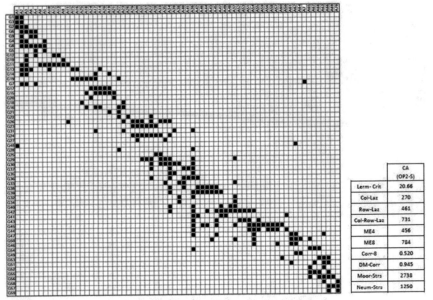

	CA (OP2-S)
Lerm- Crit	20.66
Col-Laz	270
Row-Laz	461
Col-Row-Laz	731
ME4	456
ME8	784
Corr-8	0.520
DM-Corr	0.945
Moor-Strs	2738
Neum-Strs	1250

Order on the columns obtained by the OP2-S algorithm, starting with column A1 deduced from CA and MDS-HS representations and order on the rows driven by the order on the

Fig. 4.70 OP2-S(A1) from common origin of CA and MDS-HS graphical representations

The ROFCO algorithm proves to be a very efficient tool for ordering the rows from the order of the columns and this, to obtain a seriation profile of the incidence data table. Mostly, the result obtained is more stressed than that for which the method used for ordering the rows is the same as that, for ordering the columns. As said, due to its combinatorial and statistical nature, this algorithm will be formally expressed in Chap. 5, Sect. 5.2.

Also, OP2-S algorithm proves its efficiency for ordering all the columns, from an initial column point detected visually in planar geometric representation provided by a data analysis method (PCA, CA, APS, MDS-HS) (Fig. 4.69).

OP2-S (see 5.3.1.2 of Chap. 5) was the starting point for a whole family of chaining algorithms that we will develop in the next chapter (Fig. 4.70).

References

1. J.E. Atkins, E.G. Boman, B. Hendrickson, A spectral algorithm for seriation and the consecutive ones problem. SIAM J. Comput. **28**, Nř 1, 297–310 (1998)
2. J.B. Kruskal, M. Wish, *Multidimensional Scaling* (Sage Publications, 1984)
3. J.-P. Benzécri, *L'analyse des données, tome II* (Dunod, 1973)
4. L. Lebart, J.-P. Fenelon, *Statistique et Informatique Appliquées* (Dunod, 1973)
5. I.C. Lerman, *Classification et analyse ordinale des données* (Dunod, 1981), http://www.brclasssoc.org.uk/books/index.html
6. I.C. Lerman, *Foundations and Methods in Combinatorial and Statistical Data Analysis and Clustering* (Springer Nature, 2016)
7. H. Hotelling, Analysis of a complex of statistical variables into principal components. J. Educ. Psychol. **24**(6), 417–441 (1933)
8. R. Sibson, Multidimensional scaling in theory and practice, in séminaire du Laboratoire d'Informatique pour les Sciences de l'Homme 1, editor, *Raisonnement et méthodes mathématiques en archéologie* (CNRS, 1977), pp. 73–97
9. S. Niermann, Optimizing the ordering of tables with evolutionary computation. Am. Stat. Assoc. **59**, 41–46 (2005)
10. P. Perin, La datation des tombes mérovingiennes. historiques, méthodes, applications. Paris, DROZ (1980)
11. H. Leredde, P. Perin, Les plaques boucles mérovingiennes. Dossiers de l'Archéologie **42**, 83–87, Mars-Avril (1980)
12. H. Leredde, La méthode des poles d'attraction, La méthode des poles d'agrégation. Ph.D. thesis, Université de Paris 6, October 1979
13. J. Bertin, Traitements graphiques et mathématiques. différence fondamentale et complémentarité. Mathématiques Sci. Hum. **80**, 60–71 (1980)
14. P. Ihm, A contribution to the history of seriation in archaeology, in *Studies in Classification, Data Analysis and Knowledge Organization, Proceedings of the 28th Annual Conference of the Gesellschaft für Klassifikation*, ed. by W. Gaul, C. Weihs, March 9–11, 2004, pp. 307–316. University of Dortmund (2005)
15. F.R. Hodson, *The La Tene Cemetry at Münsingen-Rain: Catalogue and relative Chronology*, vol. 50 (Berne: Verlag Stämpfil, 1968)
16. D.G. Kendall, Seriation from abundance matrices, in *Mathematics in Archaeological and Historical Sciences* ed. by D.G. Kendall, F.R. Hodson, P. Tautu (Aldine-Atherton, Chicago, 1971), pp. 214–252
17. J-P. Benzécri, *Pratique de l'Analyse des Données* (Dunod, 1980)
18. M. Hill, Correspondence analysis: a neglected multivariate method. Appl. Stat. **23**, 340–354 (1974)

Chapter 5
A New Family of Combinatorial Algorithms in Seriation

5.1 Introduction: The Fundamental Principles

In Sect. 3.6 of Chap. 3 the general idea of a new family of combinatorial algorithms was given. In this chapter we shall develop and illustrate the respective methods of this family. The general expression of the data concerned by these methods is defined by a finite set E provided with a numerical proximity matrix (similarity, dissimilarity or distance matrices), the latter being symmetrical. To fix ideas, let us put oneself into the data structure omnipresent in this book. This is defined by an incidence data table $\mathcal{T}$ crossing a set of objects $\mathcal{O}$ with a set of Boolean attributes $\mathcal{A}$ (see (2.1) and (2.2) in Chap. 2). In this case E may be the object set $\mathcal{O}$ or the attribute set $\mathcal{A}$.

What is needed is a pair of seriations which correspond each other on $\mathcal{O}$ and $\mathcal{A}$, respectively. The result is visualized by ordering the columns and rows of the incidence data table according to the seriation pair. Clearly, the same algorithm can be used to order both the column set and the row set of the data table. In the first case, the similarity index used is that expressed in Eq. (2.9) which is reduced to Eq. (2.13), the latter being reconsidered in Eq. (6.9) of Chap. 6. In the second case (ordering the row set), the proximity index is defined by Eq. (6.10) of Chap. 6.

As pointed several times, our approach is to first to establish a seriation on $\mathcal{A}$,— represented by column attributes of $\mathcal{T}$—then, to derive a seriation on the object set $\mathcal{O}$, represented by the row objects of $\mathcal{T}$. Thereby, the development of this chapter is situated within the framework of ordering the column attributes. Ordering the rows is derived from the combinatorial ROFCO algorithm (see Sect. 5.2), the latter being already introduced and used several times in the preceding chapter (see Chap. 4). Due to the importance of this algorithm, it will detailed first.

In the algorithms proposed in Sect. 5.3, we begin by defining an exhaustive set of ways for starting the seriation. For a given one, the matter consists of establishing a chaining method in order to determine the seriation, that is, an ordered sequence of all the elements of the set concerned. Denote by W this set of ways. Thereby, as already illustrated in Sect. 3.6, two stages are included in the definition of a given algorithmic.

1. Choice of an initial element among an exhaustive set of possibilities, to start the seriation;
2. Applying a chaining method to repair a new element in the subset not yet seriated in order to be aggregated to the ordered seriation section already established.

To be clear—as it was in Sect. 3.6 of Chap. 3—let us consider once more a simple case for which the attraction pole is a single element of $\mathcal{A}$. In these conditions, denote by j_1 the initial element chosen and designate by

$$(j_1, j_2, ..., j_r) \tag{5.1}$$

the ordered seriation section already built $(r < p)$.

Let

$$\mathbb{J} = \{1, 2, ..., j, ..., p\} \tag{5.2}$$

encode the attribute set (see (2.1)). j_{r+1} is the element of

$$\mathbb{K}_r = \mathbb{J} - \{j_1, j_2, ..., j_r\} \tag{5.3}$$

whose association with the ordered sequence (5.1) is the strongest.

Association coefficients between the column attribute represented by j_{r+1} and the sequence of column attributes corresponding to $(j_1, j_2, ..., j_r)$ will be proposed. In the case for which the order of the components is considered as immaterial, classical distance indices between disjoint subsets (as proposed in ascendant hierarchical clustering) can be defined (see Sect. 5.3.1).

Original and relevant adaptation of such indices will be proposed in the case where the order of the elements in the sequence $(j_1, j_2, ..., j_r)$ is taken into account (see Sect. 5.3.2.1). In this case, it is of importance that the respective proximity values of j_{r+1} with $j_r, j_{r-1}, ..., j_2$ and j_1 have to be weighted with a decreasing sequence of numerical positive weights.

Now, the most difficult is to make clear the meaning of the first item above. The general idea consists of starting the seriation from the different elements of a set of ways, the most diversified as possible and to choose the best of them.

The easiest and the most direct procedure consists of considering the set of initializations as defined by the whole set $\mathcal{A}$. In other terms, $p = card(\mathcal{A})$ initializations of the seriation have to be experimented, where the jth one corresponds to the jth Boolean attribute a^j, represented by the jth column attribute x^j (see (2.6)), $1 \leq j \leq p$. It is this type of technique which is considered in Sect. 5.3.1.

However, relative to the latter approach we may be wondering about its robustness. This because *all the seriation construction* is based on the choice of a *single element*. An alternative consists of initializing the seriation with an arrangement of h elements (h column attributes in our case) among the p elements. h is a parameter fixed by the practitioner. For computational complexity reasons h has to be small enough. This arrangement is defined by a permutation of a subset of h elements. Then, a possibility for initializing the seriation corresponds to a permutation (ordered sequence) taking the form $(j_1, j_2, ..., j_h)$ where $j_1, j_2, ..., j_{h-1}$ and j_h are mutually distinct elements in the set $\{1, 2, ..., p\}$, indicating h column attributes among the p ones (see Sect. 5.3.2). In these conditions, the initialization possibilities space includes $p \times (p-1) \times ... \times (p-h-1)$ elements.

In the case where h is not small enough, the size of the initialization possibilities space may become too large. Heuristics optimizing the search of arrangements driving seriations are proposed in Sects. 5.3.2.4 and 5.3.2.5. The respective positions of such arrangements may move along the ordered sequence defining the seriation.

To summarize. We consider two main cases to start the process of the seriation construction. In the first, we initialize with a single element (see Sect. 5.3.1). In the second case, the initialization is performed with an arrangement of h elements ($h > 1$) (see Sect. 5.3.2). In the first case, the order of the elements aggregated in the construction of the seriation may or may not intervene. In the second case, this order intervenes systematically.

The global statistical criterion (3.83), detailed in Sect. 3.3.3 of Chap. 3, can be considered for the quality evaluation of a seriation. It is defined in the case where the data is a similarity, dissimilarity or distance matrix, on the set concerned. Therefore, this type of data corresponds to a *two-way one-mode data* [1]. Otherwise, nine criteria were proposed in Sect. 3.3.4 of Chap. 3, associated with the version *two-way two-mode data*. Indeed, these criteria are defined at the level of the incidence data table (see (2.2)) of Sect. 2.1 of Chap. 2. Let us recall here the respective names and associated acronyms of the criteria:

Column Lazarus (*Col-Laz*), Row Lazarus (*Row-Laz*), Column and Row Lazarus (*Col-Row-Laz*), Measure of Effectiveness 4 (*ME4*), Measure of Effectiveness 8 (*ME8*), Correlation Coefficient according to Moore neighboring (*Corr-8*), Neumann Stress (*Neum-Strs*), Moore Stress (*Moor-Strs*) and Deutsh and Martin Correlation Coefficient (*DM-Corr*).

5.2 Row Data Table Ranking Associated with a Column Ordering in Seriation

5.2.1 Introduction

An incidence data table $\mathcal{T}$ (see (2.2)) is given such that there exists a column and row permutations revealing a σ form (see Fig. 2.1). According to σ, the column permutation is supposed here to have been already obtained. In these conditions, the problem consists of deriving from the column permutation the associated row permutation according to the σ form. This, transposed in the framework of archeology, corresponds to deduce the chronological order on the tombs from the ritual ranking.

To begin the process, take up again the formal expression of the function c (see (3.5) and what follows) defining a σ form. c is a mapping of the column set into the row set of the incidence data table. $\mathbb{J}$ and $\mathbb{I}$ designating respectively these two sets, we have

$$\mathbb{J} \longrightarrow \mathbb{I}$$
$$j \mapsto c(j) \tag{5.4}$$

$c(j)$ was denoted by $i_1(j)$ in (2.5) (see what follows this equation). $c(j)$ is the row index where the *vertical* j is loaded—in each of its cells—with the one value. In the column j, apart from the segment loaded with the one value, the cells are filled exclusively with the zero value (see Fig. 2.1). For any j in $\mathbb{J}$, the length (number of elements) of the vertical segment loaded with the one value is supposed to be constant. This length was denoted by t in (2.5). Let us take up again here the latter equation:

$$(\forall j \in \mathbb{J}), (\forall i \in \mathbb{I}), i < c(j) \text{ or } i > c(j) + t - 1, x_i^j = 0 \text{ or not defined}$$
$$\text{and}$$
$$(\forall j \in \mathbb{J}), (\forall i \in \mathbb{I}), c(j) \le i \le c(j) + t - 1, x_i^j = 1 \tag{5.5}$$

Now, let us consider the formal row description of an incidence data table, underlined by a σ form. This was given in (2.4). A more precise version using the above mapping c will be proposed here. Every row of the incidence data table includes an *interval* loaded with the one value in each of its cells. Apart from this interval all the values contained in the row concerned are equal to 0. In the following this interval is referred as the "loaded interval". If i is the index of a given row ($i \in \mathbb{I}$), the rank in $\mathbb{J}$ of the right boundary of its loaded interval is equal to $c^{-1}(i)$. Designating by $l(i)$ the length of the loaded interval, we have for $i \le p$ (see Fig. 2.1):

$$(\forall j \in \mathbb{J}), (\forall i \in \mathbb{I}), x_i^j = 1 \text{ if } c^{-1}(i) - l(i) + 1 \le j \le c^{-1}(i)$$
$$\text{while for } i > p$$
$$x_i^j = 1 \text{ if } p - l(i) \le j \le p \tag{5.6}$$

Now, as indicated above, we assume that the column ordering of the incidence data table $\mathcal{T}$ corresponds to a σ form underlying $\mathcal{T}$. However, the row ordering is not supposed to follow necessarily the σ form. That is, the rows of the data table are supposed permuted in any way (arbitrarily) with respect to σ (see Table 5.1). This circumstance might be after applying a method which only enables us to order the column attributes. The property defined by (5.6) holds in the latter case. In other terms, the set of cells of a given row, loaded with the one value, constitutes an interval of the column sequence. Taking into account this property, the problem posed is to retrieve the serial order on the row set from that on the column set.

Table 5.1 Sigma form with disorganized rows

	1	2	3	4	5	6	7
1	X	X					
2			X	X			
3						X	X
4							X
5	X						
6				X	X		
7		X	X				
8					X	X	

Table 5.2 Sigma form with ordered rows

	1	2	3	4	5	6	7
5	X						
1	X	X					
7		X	X				
2			X	X			
6				X	X		
8					X	X	
3						X	X
4							X

With this end in view, two algorithms will be proposed. The first one, called "Combinatorial Algorithm" (see Sect. 5.2.2), has a pure formal character and in the second one, called "Statistical Algorithm", each row is interpreted as a statistical distribution on the column set (see Sect. 5.2.3).

Let us point out here that fuzzy σ forms were randomly simulated in [2]. For the latter—by difference with an exact σ mathematical form, a value one (resp., a value zero) is replaced by the result of a random binary simulation for which a one (resp., a zero) value has with respect to the zero (resp., to the one) value, a high probability to occur. For such fuzzy forms—resulting from random simulations—the "Statistical Algorithm" has to be employed (Table 5.2).

5.2.2 The Combinatorial Algorithm

Consider up again the rectangular incidence data table whose dimension is $n \times p$ (n rows and p columns). n is the object number and p, the attribute number. We suppose that a σ form underlies the data table structure. On the other hand, the column attributes are supposed to have been ordered from 1 to p, according to the latter structure. An illustration is given in Table 5.1.

As just expressed above, according to (5.6), whatever the ordering on the set of the row objects, in each row there is a loaded interval, that is to say, an interval such that each of its cells includes the one value. From both boundaries of the latter interval, on the left and on the right, all of the cells contain a zero value. Notice that is exactly one row for which the interval concerned has its first (resp., last) element at column 1 (resp., p). As emphasizing above, the problem consists of discovering the row serial order as it appears in Fig. 2.1.

For this purpose two order properties on the row set will be defined and exploited. To begin, we associate with the ordered column sequence from 1 to p, an ordinal axis oriented from left to right. To a loaded interval of a given row corresponds the abscissa of its first element, the latter may be 1, 2, ..., $p - 1$ or p. For the reconstructed σ form, this abscissa is an increasing function of the ordering to establish on the rows (see Fig. 2.1). It corresponds to the beginning of the loaded interval. This states the first of the two order properties mentioned.

On the other hand, we observe that for the row ordering sought (see Fig. 2.1), the length (number of elements of the loaded interval) is first increasing and next decreasing, the increase (resp., decrease) not being strict. This defines the second order property.

In these conditions, a total preorder resulting from the lexicographic product of the two total preorders, respectively associated with the preceding order properties, is defined. More explicitly, the first total preorder, for a given row i, is defined according to the range $l(i)$ of the loaded interval (see (5.6)) and the second one is based on the smallest of the two following distances:

1. Distance between the abscissa of the start of the ith row loaded interval and the left boundary of the row concerned, this distance can be designated by $\delta_g(i)$;
2. Distance between the abscissa of the end of the ith row loaded interval and the right boundary of the row concerned, this distance can be designated by $\delta_d(i)$.

Therefore,

$$(\forall i \in \mathbb{I}), l(i) = p - [\delta_g(i) + \delta_d(i)] \tag{5.7}$$

As illustration, we have in the case of Table 5.1:

$$\delta_g(7) = 1 \text{ and } \delta_d(7) = 4$$

$$\delta_g(5) = 0 \text{ and } \delta_d(5) = 6$$

And then,

$$l(7) = 7 - (1 + 4) = 2$$

$$l(5) = 7 - (0 + 6) = 1$$

Now, for each of the preorder classes, the i row elements belonging to it are ordered according to the value of

$$min[\delta_g(i), \delta_d(i)] \qquad (5.8)$$

And then, for each of them

1. If in (5.8) the lowest distance is $\delta_g(i)$, then the row concerned is piled up;
2. If in (5.8) the lowest distance is $\delta_d(i)$, then the row concerned is stacked downward.

Relative to the example provided by Table 5.1 the first total preorder is given by

$$4 \sim 5 < 1 \sim 2 \sim 3 \sim 6 \sim 7 \sim 8 \qquad (5.9)$$

where 1, 2, 3, 4, 5, 6, 7 and 8 are the row labels and where $\sim$ denotes the equivalence binary relation induced by the partition associated with the total preorder.

Consequently, with suggestive notations, the piling up order is obtained as follows:

$4 \rightarrow$ down, $5 \rightarrow$ top, $1 \rightarrow$ top, $3 \rightarrow$ down, $7 \rightarrow$ top, $8 \rightarrow$ down, $2 \rightarrow$ top, $6 \rightarrow$ down

In these conditions, the reorganization of the data table rows is carried out according to Table 5.1, where the serial σ form appears clearly.

5.2.3 The Statistical Algorithm

The indexing set $\mathbb{J} = \{1, 2, ..., m, ..., p\}$ of the column sequence of the data table is interpreted here as the interval $[1, 2, ..., m, ..., p]$ of the integers. In the case of a σ form of the incidence data table (see Fig. 2.1), the loaded interval of a given row is considered as an interval of $[1, 2, ..., m, ..., p]$. This property is satisfied whatever is the row ranking. As an example in Table 5.1, the loaded interval of the row 2 is [3, 4], that of the row 5, is the interval [1], reduced to a single element.

Now, relative to the loaded interval of a given row, we associate a probability distribution on $[1, 2, ..., m, ..., p]$ defined as follows.

Let $I = [l, m]$ designate the latter loaded interval, the random variable K defining the probability distribution can be expressed by:

$$Pr\{K = k\} = \begin{cases} 0 & \text{if } k < l \text{ or } k > m \\ 1/m - l + 1 & \text{if } l \leq k \leq m \end{cases}$$

In these condition, if $I' = [l', m']$ is, for a given row a loaded interval of the same length as I ($m' - l' + 1 = m - l + 1$), for which $l < l'$, then the random variable K' associated with I', is stochastically on the right of K. In more formal words

$$(\forall h, l \leq h \leq m'), \; Pr\{K \leq h\} \geq Pr\{K' \leq h\} \tag{5.10}$$

As a consequence, the mean of K is strictly smaller than that of K'. Therefore, the center of the K distribution law, which is the middle of I, is on the left of the center of the K' distribution law, the latter being the middle of I'. Thereby, by considering in the example of Table 5.1, the loaded intervals $I = [2, 3]$ and $I' = [3, 4]$, respectively, associated with the rows 7 and 2, the distribution of K is $\{(2, 0.5), (3, 0.5)\}$ and that of K' is $\{(3, 0.5), (4, 0.5)\}$. The center of the first distribution is

$$0.5 \times 2 + 0.5 \times 3 = 2.5$$

and that of the second distribution

$$0.5 \times 3 + 0.5 \times 4 = 3.5$$

In these conditions, the row 7 will be placed before the row 2.

Let us now consider the case where the intervals I and I' are of inequal lengths. Without generality restriction, imagine that the interval I starts before than I'. In the case of the seriation structure as considered, the end of the interval I occurs before than that of I'. Both intervals may either have a non-empty intersection or be disjoint. Thus, for example, in Table 5.1, by considering the I interval of row 8 and the I' interval of row 4, I ends strictly before than I' starts. Besides, if we consider the interval I as associated with the row 3 and that I', with the row 4, I' and I end simultaneously. In these conditions, the property defined by (5.10) holds. Therefore, we have to put first on the top, the row for which the center of its loaded interval has the smallest abscissa. To conclude, the rows have to be arranged according the abscissa of the middle of their respective loaded intervals.

In the case of the example given by Table 5.1, the row ordered sequence we obtain is $(5, 1, 7, 2, 6, 8, 3, 4)$. And then, the result obtained by the "Combinatorial Algorithm" is come back exactly. However, as mentioned above at the end of the Introduction section, the "Statistical Algorithm" is much more flexible than the combinatorial one. Consequently, the "Statistical Algorithm" has to be used in real cases where the σ form associated with a seriation statistically dominates without being verified exactly mathematically.

As said several times this algorithm is designated by the acronym **ROFCO** which symbolizes **R**ow **O**rder **F**rom **C**olumn **O**rder. It was particularly used in Chap. 4 where from a planar geometric representation of all the columns, we could infer a seriation order on them. This order on the columns leads via ROFCO to an order on the rows corresponding to the seriation.

It may be interesting to study the variation in the behavior of a method after removing certain rows from the incidence table. ROFCO offers the possibility of reinjecting the extracted rows relative to the seriation structure obtained without these last rows. This was done in relation to the PCA method in the case of the Münsingen-Rain data table. The rows extracted are those comprising the smallest

number of attributes present (they were two attributes present). The results have been
significantly improved.

5.3   Seriation Algorithms

5.3.1   The Seriation Search Space Is Associated with the First Element

General considerations

Let us resume some points of the introduction above (see Sect. 5.1). The set
to be seriated first is the set $\mathcal{A}$ of attributes. Two main stages noted 1 and 2 were
mentioned. For the first one, here, the parameter of the seriation is defined by its
first element. In this case there are p possible initializations, corresponding to the
p column attributes, respectively. For a given iteration (there are in all p), we will
designate as above, by j_1, the initialization element. Moreover, $(j_1, j_2, ..., j_r)$ will
denote the already constituted sequence of column attributes for the seriation $(r < p)$.
j_{r+1} is the element of

$$\mathbb{K}_r = \mathbb{J} - \{j_1, j_2, ..., j_r\} \tag{5.11}$$

the more associated with the ordered sequence $(j_1, j_2, ..., j_r)$ (see (5.1), (5.2) and
(5.3) above). In the case where several elements carry out simultaneously the
strongest association concerned, one option may consist of aggregating them in
an arbitrary order following j_r in the sequence $(j_1, j_2, ..., j_r)$. On the other hand,
clearly, if $r = p - 1$, j_p is aggregated following j_{p-1}.

Two alternatives can be envisaged for establishing the association coefficient
between j_{r+1} and the ordered sequence $(j_1, j_2, ..., j_r)$. For the first one, the order of
this sequence is not taken into account and for the second one, it is essential.

Recall that we have chosen to order first the attribute set $\mathcal{A}$. For this purpose, we
assume $\mathcal{A}$ provided with a distance measure as that defined in Eq. 6.11 of Chap. 6,
this being deduced from Eq. 2.13 of Chap. 2. The general expression we consider
for the distance matrix is

$$\mathcal{D}(\mathbb{J}) = \{d(k, l)|1 \leq k, l \leq p\} \tag{5.12}$$

5.3.1.1   No Role for the Order of the Already Aggregated Elements

According to the notations adopted (see above), $(j_1, j_2, ..., j_r)$ designates the ordered
sequence of the already seriation built. From this ordered sequence we only retain
the subset structure $\{j_1, j_2, ..., j_r\}$. The j_{r+1} element has to be sought in the
complementary subset $\mathbb{K}_r$ (see (5.11)) in order to obtain the ordered sequence

$(j_1, j_2, ..., j_r, j_{r+1})$. For this, each of the binary hierarchical clustering can be used (see Chap. 10 of [3]). Classical distance indices between classes (clusters) coming from ascendant hierarchical clustering can be proposed for this case. In the latter, one of the two classes to be compared, namely, $\{j_1, j_2, ..., j_r\}$, is fixed. The other one, $\{j_{r+1}\}$ is a singleton class. It corresponds to one element to be sought in $\mathbb{K}_r$ (see (5.11)). In order to illustrate this point let us recall two very classical distance indices between disjoint clusters: the maximal link ("Single Linkage") and the proximity average ("Average Linkage") that we denote by D_{sl} and D_{al}, respectively:

$$D_{sl}(j_{r+1}, \{j_1, j_2, ..., j_r\}) = min\{d(j_{r+1}, j_q)|1 \leq q \leq r\} \qquad (5.13)$$

and

$$D_{al}(j_{r+1}, \{j_1, j_2, ..., j_r\}) = min\{\frac{1}{r}.\sum d(j_{r+1}, j_q)|1 \leq q \leq r\} \qquad (5.14)$$

The respective acronyms to designate the seriation algorithms employing D_{sl} and D_{al} are **OP-SL** and **OP-AL**: **OP-SL** for **O**rdinal **P**roximity with respect to the **S**ingle **L**inkage criterion and **OP-AL** for **O**rdinal **P**roximity with respect to **A**verage **L**inkage criterion.

Considering the chaining nature of the needed association, it is the criterion of the Single Linkage which seems the most suitable and then, it will be experimented.

5.3.1.2   A Role for the Order of the Already Aggregated Elements

First Criterion of Association for the Seriation

According to our formulation above $(j_1, j_2, ..., j_r)$ indicates the already sequence defining a given step in the construction of the seriation. Now, we shall consider two distance criteria of a given element j_{r+1} of $\mathbb{K}_r$ (see (5.11)) with the latter ordered sequence. These criteria have to be minimized. For the first of them that we can call "Simple Successive Chaining" and denote by $S - Chain$, the comparison—between j_{r+1} and the beginning section already established—is performed with the last element of the latter sequence, that is to say, with j_r. This method has already been considered in Sect. 3.6 of Chap. 3, in the context of the seriation of the object set $\mathcal{O}$. Here, we have

$$S - Chain(j_{r+1}, (j_1, j_2, ..., j_r)) = d(j_{r+1}, j_r) \qquad (5.15)$$

where d is a distance measure on the attribute set (see (2.88) in Chap. 2).

The acronym used for the seriation algorithm using $S - Chain$ (see (5.15)) is **OP-LA** for **O**rdinal **P**roximity with respect to the **L**ast **A**ggregated element.

Second Criterion of Association for the Seriation

In the second association criterion that we call "Weighted Successive Chaining" and we denote by $W-Chain$, the comparison between j_{r+1} and the sequence $(j_1, j_2, ..., j_r)$ is weighted. The respective weights are numerical positive numbers, the sum of them being unity. In these conditions we have

$$W-Chain\big(j_{r+1}, (j_1, j_2, ..., j_r)\big) = \sum_{1 \leq g \leq r} \alpha_k d(j_{r+1}, j_k) \qquad (5.16)$$

where

$$0 < \alpha_1 < \alpha_2 < ... < \alpha_k < ... < \alpha_r$$
and where
$$\sum_{1 \leq k \leq r} \alpha_k = 1 \qquad (5.17)$$

In our experimental analysis we have taken

$$\alpha_k = \frac{2k}{r(r+1)}, \ 1 \leq k \leq r \qquad (5.18)$$

Nevertheless, other options can be envisaged for the weights α_k.

In the latter equations ((5.15) and (5.16)) the square d^2 of the distance d can be substituted for d. Relative to an incidence data table, in the case where the set to be organized is the attribute set (resp., the object set) D_α^2 (resp., D_ω^2), as defined in Eq. (6.38) (resp., (6.39)) of Sect. 6.2 of Chap. 6 can be employed.

Now, these equations ((5.15) and (5.16)) can also be considered as they are, in the case where a similarity coefficient is used instead of a distance index. More explicitly, relative to an incidence data table, in the case where the set to be organized is the attribute set (resp., the object set), S (resp., P) as defined in (6.36) (resp., (6.37)) of Sect. 6.2 of Chap. 6 can be used. However, the criteria defined by (5.15) and (5.16) have to be maximized and not minimized.

This method is designated with the acronym **OP2-S**. This stands for **O**rdinal **P**rogressive **P**roximity with respect to a **S**ingle element. The progressive nature of the proximity is directly related to the weighting as defined in equations (5.16) to (5.18). In this acronym **P2** refers to Progressive Proximity and **S** to Single element.

Computational Complexity

The number of point distances computed or consulted (in the case where the distance matrix is established) in order to determine the seriation with the first association criterion ("$S - Chain$") can be put in the following form

$$p \times [(p-1) + (p-2) + \cdots + 2 + 1] = \frac{1}{2}(p-1) \times p^2 \qquad (5.19)$$

Considering the left member, the first factor p corresponds to the p possible alternatives for j_1. In this expression we consider that the last distance considered is between j_p and j_{p-1}.

Now, let us specify the number of distances computed or consulted in order to establish the seriation with the second association criterion ("$W-Chain$"). It can be expressed as follows:

$$p \times \{(p-1) \ \times 1 + (p-2) \times 2 + \cdots + [p-(p-2)]$$
$$\times (p-2) + [p-(p-1)] \times (p-1)\} \qquad (5.20)$$

In the latter equation—as above—the first factor p corresponds to the p possible alternatives for the choice of j_1. In this expression, the last distances considered are between j_p and the respective elements of the sequence $(j_1, j_2, ..., j_{p-1})$. This equation can be reduced to

$$p \times \{p \times \frac{p(p-1)}{2} - \frac{p(p+1)(2p+1)}{6}\} = \frac{p^2}{6} \times (p^2 - 6p - 1) \qquad (5.21)$$

5.3.2   The Seriation Search Space Is Associated with Sized Sequences of Elements

5.3.2.1   Introduction

The nature of the seriation problem considered in this chapter refers to combinatorial optimization [1]. Therefore, its computational complexity is Non-Deterministic Polynomial. In these conditions, heuristics are needed in order to approximate the solution.

In the previous section (see Sect. 5.3.1) the cardinal of the search space is p, corresponding to the p elements of the set to be ordered, $\mathcal{A}$. Here, we shall considerably enlarge this space. Each of its elements is an arrangement of $\mathcal{A}$ elements sized by the integer h. The h value has to be small enough in order to make manageable the computational complexity. An h-arrangement is defined by an ordered sequence of h mutually distinct elements. It can be obtained by strictly ordering a subset H of $\mathcal{A}$, h sized. In fact, for a given h subset H, all the arrangements derived from it are considered. The total number of these arrangements is

$$p^{[h]} = \binom{p}{h} \times h! = p \times (p-1) \times (p-2) \times ... \times (p-h+1) \qquad (5.22)$$

In effect, $\mathcal{A}$ being encoded with $\{1, 2, ..., j, ..., p\}$ each h subset of the latter ($h < p$) gives rise to $h!$ permutations and each of these determines an arrangement of the subset concerned.

For illustration, consider $p = 24$ and $h = 3$. The arrangements associated with the subset $\{12, 15, 22\}$ are

$$(12, 15, 22), (12, 22, 15), (15, 12, 22), (15, 22, 12), (22, 12, 15), (22, 15, 12)$$

In this section the search space parameter is not reduced to a single element of $\mathcal{A}$, but to an h subset of $\mathcal{A}$ or an ordered sequence of h subsets. The unique h subset or the first one of the latter sequence is associated with the start of the seriation construction.

In fact, as just expressed, what is handled is the set of h—arrangements derived from an h—subset. A given h—arrangement situated at a given rank position and participating in the seriation construction will be considered as an *active h—window*. More precisely, an active h-window takes part in the construction of a seriation, by adding elements not yet seriated.

Now and before to go ahead, let us indicate some features concerning the algorithms developed in the following.

In Sect. 5.3.2.2, entitled OP3-FIA algorithm, only one h-window placed at the beginning of the seriation participates to its construction. In Sects. 5.3.2.3, 5.3.2.4, 5.3.2.5, the seriation construction is associated with an ordered sequence of active h-windows.

In Sect. 5.3.2.3, the called OP3-SAA algorithm is expressed. It corresponds to an iterated sequence of the preceding algorithm. At a given stage, the data is defined by a series of arrangements sized by h and mutually disjoint. Note that two arrangements $(x_1, x_2, ..., x_h)$ and $(y_1, y_2, ..., y_h)$ are defined as disjoint, if it is the case for the sets concerned, namely,

$$\{x_1, x_2, ..., x_h\} \cap \{y1, y_2, ..., y_h\} = \emptyset$$

In Sect. 5.3.2.4, entitled OP3-OSA algorithm, the active h-window considered is at the extreme right of the start of the seriation already constructed. If g is the length of the latter ($g \leq p - 2$), the positions occupied by the active window are from $(g - h + 1)$ to g. The first of them—defining the start of the seriation—depends on the choice of a single element. This determines the first h-window concerned by using an ordinal proximity notion. All of the elements—there are p—are tested in this respect.

In Sect. 5.3.2.5, entitled OP3-SSA algorithm, by denoting g the length of the seriation start already constructed, there are $(g - h + 1)$ h-windows in all. The positions occupied by the kth one are from k to $(k - h + 1)$, $1 \leq k \leq (g - h + 1)$. If x is an element not yet seriated, it occurs in $(g + h - 1)$ comparisons, with respect to the sequence already seriated. Each of which is permutational for one of the h-windows. The comparison assumes progressive proximity as defined in Sect. 5.3.1.2.

5.3.2.2   Development of OP3-FIA Algorithm

OP3-FIA is an acronym associated with **O**rdinal **P**rogressive and **P**ermutational **P**roximity with **F**ixed **I**nitial **A**rrangement.

In this algorithm the attraction pole notion, represented by a single element of $\mathcal{A}$, as considered in Sect. 5.3.1.2 is replaced by an arrangement of h elements. The latter is obtained from a subset H of $\mathcal{A}$ including h elements. In fact, there are $h!$ arrangements associated with a given subset H of h elements. And, there are in all $p^{[h]}$ arrangements (see (5.22)). As already said, h has to be small enough for computational complexity reasons.

To each element of these arrangements we will associate by "Ordinal, Progressive and Permutational Proximity" (Π), a unique seriation.

In clear, denote by

$$\sigma_0 = (j_1^0, j_2^0, ..., j_h^0) \tag{5.23}$$

a given initial arrangement $(j_1, j_2, ..., j_h)$. This defines the start of a given seriation. Following the first increment, the beginning of the seriation becomes

$$(j_1^1, j_2^1, ..., j_h^1, j_{h+1}^1) \tag{5.24}$$

where

$$(j_1^1, j_2^1, ..., j_h^1) = (j_1^0, j_2^0, ..., j_h^0)$$

and where j_{h+1}^1 is obtained by the formula

$$j_{h+1}^1 = Arg\{min\{ \sum_{1 \leq k \leq h} \alpha_k d(j, j_k^0) | j \in K\}\} \tag{5.25}$$

where the weights α_k, $1 \leq k \leq h$, are defined in (5.18) and where K is the complementary subset of H. The interval corresponding to the h first positions of the seriation can be called h active window.

After getting j_{h+1}^1 from (5.25), (5.24) becomes

$$(j_1^1, j_2^1, ..., j_h^1, j_{h+1}^1) = (j_1^0, j_2^0, ..., j_h^0, j_{h+1}^1) \tag{5.26}$$

Thereby, the h-arrangement which initializes the seriation is fixed with respect to the sequence of the elements achieving the seriation.

In words, the optimization is relative to an ordered pair where the first argument is an element of K and where the second one is an arrangement as defined by (5.23). However, in the result retained (see (5.26)), the h first components are preserved according to the arrangement considered in order to initialize the seriation.

More generally and more precisely, let us consider the step for which the $(l + 1)$th element has to be determined, $1 \leq l \leq p - h - 2$. The already built sequence can be indicated as follows:

$$(j_1^l, j_2^l, \ldots, j_{h+l-1}^l, j_{h+l}^l) \tag{5.27}$$

where

$$(j_1^l, j_2^l, \ldots, j_h^l) = (j_1^0, j_2^0, \ldots, j_h^0)$$

(see (5.23), and where

$$j_{h+1}^l = j_{h+1}^1, \, j_{h+2}^l = j_{h+2}^2, \, \ldots, \, j_{h+l-1}^l = j_{h+l-1}^{l-1}, \, j_{h+l}^l \tag{5.28}$$

Following (5.28), let us define the rule enabling the ordered sequence

$$(j_1^{l+1}, j_2^{l+1}, \ldots, j_h^{l+1}, j_{h+l}^{l+1}, \ldots, j_{h+l+1}^{l+1}) \tag{5.29}$$

Notice that the upperscript is defined by the rank of the last element aggregated to the in construction seriation. In (5.29), j_{h+l+1}^{l+1} is the $(l+1)$th element which has joined the seriation in construction. $h+l+1$ is the length of the obtained seriation, the latter starting with the arrangement (5.23).

Now, let us specify the penultimate association:

$$(j_1^{p-1-h}, \ldots, j_h^{p-1-h}, j_{h+1}^{p-1-h}, \ldots, j_{h+(p-1-h)}^{p-1-h}) \tag{5.30}$$

and the last aggregation with the last element:

$$(j_1^{p-h}, \ldots, j_h^{p-h}, j_{h+1}^{p-h}, \ldots, j_{h+(p-h)}^{p-h}) \tag{5.31}$$

The determination of j_{h+l+1}^{l+1} is carried out according to

$$j_{h+l+1}^{l+1} = Arg\{min\{ \sum_{1 \leq g \leq h+l} \alpha_k.d(j, j_k^l) | j \in K_{h+l}\}\} \tag{5.32}$$

where α_k is defined in (5.18) with $r = h + l$ and where K_{h+l} is the remainder (complement) subset of

$$H_{h+l} = \{j_1^l, j_2^l, \ldots, j_h^l, j_{h+1}^l, j_{h+2}^l, \ldots, j_{h+l}^l\} \tag{5.33}$$

in $\{1, 2, \ldots, p\}$.

The reader interested can easily transpose the equations (5.25) to (5.32) in the case of referring to a similarity matrix of the form

$$S(\mathbb{J}) = \{S(k, l) | 1 \leq k, l \leq p\} \tag{5.34}$$

instead of the dissimilarity matrix $\mathcal{D}(\mathbb{J})$ (see (5.12)).

Number of times an element of $\mathcal{D}(\mathbb{J})$ takes part

Consider the passage from (5.27) to (5.28). This uses Eq. (5.32) above. The latter includes a sum defining the value of the comparison of the element to be aggregated— that we have to seek in the remainder subset $\{j | h + l + 1 \leq j \leq p\}$, with each element of the ordered section going from 1 to $(h + l)$ in (5.27).

An h arrangement is defined by an h subset and a permutation on its elements. For a given initial arrangement there are $[p - (h + l)] \times (h + l)$ point distances which participate in order to determine j_{h+l+1}^{l+1} from

$$(j_1^l, j_2^l, ..., j_h^l, ..., j_{h+l}^l)$$

By varying the permutation on the first h elements

$$h! \times [p - (h + l)] \times (h + l)$$

point distances have to be integrated.

Finally, by varying the h subset, this gives

$$h! \times [p - (h + l)] \times h$$

point distances to be integrated. And finally, by varying the h subset, this gives in all

$$\binom{p}{h} \times h! \times [p - (h + l)] \times (h + l) = p^{[h]} \times [p - (h + l)] \times (h + l) \quad (5.35)$$

point distances in A ($p^{[h]} = p(p - 1)(p - 2)...(p - h + 1)$).

Thus, the number of times where elements of the matrix $\mathcal{D}(\mathbb{J})$ occur is

$$p^{[h]} \times \sum_{0 \leq l \leq (p-h-1)} [p(h + l) - (h + l)^2] \quad (5.36)$$

By considering $(h + l)$ as variable summation, the development of this equation gives, up to the multiplicative factor $p^{[h]}$,

$$\{p \times [\frac{p(p-1)}{2} - \frac{h(h-1)}{2}] - [\frac{(p-1)p(2p-1)}{6} - \frac{(h-1)h(2h-1)}{6}]\}$$
$$= \{\frac{1}{2}p[p(p - 1) - h(h - 1)] - \frac{1}{6}[(p - 1)p(2p - 1) - (h - 1)h(2h - 1)]\}$$
$$(5.37)$$

Once more, let us emphasize on the principle of the method described. Each of the seriations produced is generated from an arrangement of h elements among p elements, the latter arrangement defining the beginning of the seriation concerned. Each h subset gives rise to $h!$ arrangements. As already expressed (see (5.22)), there are in all

$$p^{[h]} = \binom{p}{h} \times h! = p(p-1)...(p-h+1) \qquad (5.38)$$

arrangements.

According to our notations, the seriation associated with a given arrangement is written

$$\left(j_1^{p-h}, j_2^{p-h}, ..., j_h^{p-h}, j_{h+1}^{p-h}, ..., j_p^{p-h}\right) \qquad (5.39)$$

The final stage consists of associating with each of the seriations obtained, the value of a quality criterion. This in order to select the most efficient seriation. In fact, there are several evaluation criteria. At the end of Sect. 5.1, nine of them have been listed. A *ten*th one was developed in Sect. 3.3.2 of Chap. 3. All of the criteria were illustrated in Chap. 4. Given the nature of the problem which is a selection problem, we will limit ourselves to using the simplest criteria which are Column-Lazarus, Row Lazarus and Column and Row Lazarus.

5.3.2.3 OP3-SAA Algorithm

OP3-SAA is an acronym associated with **O**rdinal **P**rogressive and **P**ermutational **P**roximity with respect to each of a **S**equence of **A**djacent **A**rrangements. The disjunction between arrangements is defined here by the disjunction between the h subsets providing these arrangements.

According to the terminology used above an h window in the seriation is an ordered sequence of related positions in the seriation. The value of an h window is the sequence of the elements occupying these positions. Thus, the value of an h window is an arrangement.

In the OP3-FIA algorithm (see Sect. 5.3.2.2), the search parameter is a fixed h window—that we have qualified as *active*—situated at the origin of the ordered sequence defining the seriation. Here, the matter consists of mutually disjoint h windows, where two consecutive of them are adjacent.

Suppose $p = k \times h + r$, where $1 \leq r \leq h$. In these conditions, the lth h window occupies the positions

$$(l-1)h + 1, (l-1)h + 2, ..., lh \qquad (5.40)$$

$1 \leq l \leq k$.

Now, designate by A_l the subset composed of the elements intervening in the latter arrangement. We have

$$A_l \subseteq \mathcal{A} - (A_1 + A_2 + \cdots + A_{l-1}) \qquad (5.41)$$

where, recall, $\mathcal{A}$ is the whole set of the column attributes, $1 \leq l \leq k$. Note that the content of the parenthesis is empty for $l = 1$. Also, denote by A_{k+1} the subset (5.41) for $l = k + 1$. A_{k+1} includes r elements where $1 \leq r \leq h$. A_{k+1} cannot give an active h window draining the end of the seriation. The latter is obtained from the right member of (5.41), with $l = k$. In this case, the cardinality of the subset defined by the right member is $h + r$.

In these conditions, there are in all

$$p^{[h]} \times (p - h)^{[h]} \times \ldots \times [p - (g - 1)h]^{[h]} \tag{5.42}$$

series of entire sequences of g h windows, where the lth one is provided with an arrangement from $[p - (l - 1)h]$ elements corresponding to the right member of (5.41).

For a given sequence of h windows k seriations are calculated. The first one is associated with the first h window which is supposed to include a given arrangement from $\mathcal{A}$. This arrangement defines the h first elements of the seriation. The rest of the elements of the seriation are assigned one by one according to the magnitude of a progressive proximity (as in (5.32)) to the already established ordered sequence defined by the seriation in construction.

Relative to the given sequence of h windows, the lth seriation is associated with the sequence of the l first h windows. The ordered sequence of their respective compositions (each composition is an arrangement) defines the start of the seriation. The rest of it is obtained element by element from

$$\mathcal{A} - (A_1 + A_2 + \cdots + A_l) \tag{5.43}$$

according to the magnitude of a progressive proximity (as in (5.32)) to the already established ordered sequence defined by the seriation in construction.

Now, by varying in all possible ways the entire sequence of k successive h windows, we can see, according to (5.39) that

$$k \times p^{[h]} \times (p - h)^{[h]} \times \ldots \times [p - (k - 1)h]^{[h]} \tag{5.44}$$

seriations are calculated in all.

5.3.2.4 OP3-OSA Algorithm

OP3-OSA is an acronym associated with **O**rdinal **P**rogressive and **P**ermutational **P**roximity with respect to a sequence of **O**rdered **S**equences of **A**rrangements, each deduced from the preceding one by a translation of one element. Only one active h window intervenes at the extreme right. A loop whose extent is p enables the different elements of $\mathcal{A}$ to take part at the first position.

The algorithmic search is parametrized by the set $\mathcal{A}$. Each element of $\mathcal{A}$ induces a seriation and $p = card(\mathcal{A})$ seriations are generated.

A specific notion in this algorithm concerns the proximity of an element not yet seriated to the ordered section of the seriation that has just been established.

Let us make explicit the successive steps of this algorithm.

1. Starting with an element of $\mathcal{A}$ coded j_1, an h window $(j_1, j_2, ..., j_h)$—driven by j_1 is built at the extreme left of the seriation, in this construction j_k is the nearest $(j_1, j_2, ..., j_{k-1})$ according to progressive and permutational proximity in which all the permutations of $(j_1, j_2, ..., j_{k-1})$ are compared, $2 \leq k \leq h$;
2. An $(h + 1)$ element is aggregated to $(j_1, j_2, ..., j_h)$ according to the same technique as in 1, and the h window is moved one position on the right becoming $(j_2, j_3, ..., j_{h+1})$, j_1 fixed at the extreme left, becomes the first element of the seriation;
3. An $(h + 2)$ element is aggregated to $(j_1, j_2, ..., j_h, j_{h+1})$ according to a progressive and permutational proximity to the entire sequence from j_1 to j_{h+1}, the permutational aspect being relative to the sub-sequence $(j_2, ..., j_h, j_{h+1})$ defined by the h window;
4. j_2 is left directly after j_1 and (j_1, j_2) determines the start of the seriation associated with the choice of j_1 as its first element;
5. The first step is that for which the length (number of elements) of the established seriation is $(h + 1)$, the kth step is that for which the length of the established seriation is $h + k$, the active window occupying the positions $k + 1$ to $h + k$, $1 \leq k \leq p - h - 1$;
6. In the last step where a single element (the pth one) remains at the extreme right, the latter is aggregated following without any calculation.

Let us recall that in the case for which the length of the established section is $g = h + k$, the respective positions of the active window go from $(k + 1)$ to $(k + h)$, $1 \leq k \leq (p - h - 2)$.

After instigating the first element j_1 the number of times where a point distance participates in the calculation is

$$1!(p - 1)! + 2!(p - 2)! + \cdots + (h - 1)!(p - h + 1)$$
$$+ (p - h) \times h! + [p - (h + 1)] \times (h! + 1) + [p - (h + 2)] \times (h! + 2)$$
$$+ \cdots + 2 \times \left(h! + (p - 2 - h)\right) \quad (5.45)$$

Note that the last factor 2 corresponds to $[p - (p - 2)]$. Otherwise, the first row of this equation is associated with the respective distances for calculating the h-window starting the seriation.

5.3.2.5 OP3-SSA Algorithm

OP3-SSA is an acronym associated with **O**rdinal **P**rogressive and **P**ermutational **P**roximity with respect to a **S**equence of **S**uccessive **A**rrangements where the last one is deduced from the preceding one by adding one element at the extreme right and by subtracting one element at the extreme left.

This algorithm has to be compared with the preceding one (see Sect. 5.3.2.4). In the latter, the progressive and permutational proximity notion uses a unique active h window at the extreme right of the established seriation section for seeking one additional element. Here, we use a series of active h windows, each of which is deduced from the previous one by adding an element on the right and subtracting an element on the left.

The first h window is determined as for the previous algorithm OP3-OSA. Let us denote by g the already acquired length of the beginning section of the seriation. For $g < h$, the $(g + 1)$th element is among the remaining (p - g) elements the one for which the permutational proximity to the sequence of the g elements is the greatest. The number of comparisons to reach this first h-window is

$$(p - 1) + (p - 2) \times 2 \times 2! \ + (p - 3) \times 3 \times 3! + \cdots$$
$$+ (p - h + 1) \times (h - 1) \times (h - 1)! \quad (5.46)$$

The change from the value of g equal to h to that of $h + 1$ is treated in exactly the same way. Things change right after.

Let us imagine we have reached a value of g of the following form

$$g = h + k \text{ where } k \geq 0$$

and consider the addition of the $(g + 1)$th element.

The number of the remaining elements, on the right of the established start of the seriation, is

$$p - g = p - (h + k)$$

In these conditions, k active h windows are considered:

$$1 \text{ to } h, 2 \text{ to } h + 1, 3 \text{ to } h + 2, \ldots, k \text{ to } (h + k)$$

Now, designate by Rem-Set the remaining set of the elements not yet seriated. $p - (h + k)$ is the cardinality of this set. Each element of the latter is compared using permutational proximity with each of the active windows. The aptitude to be seriated of a given element of Rem-Set is measured by the highest proximity. The element for which this aptitude is strongest is chosen to be added to the seriation in construction.

The number of similarity or distance indices between two elements calculated in order to determine the element to be seriated is

$$[p - (h + k)] \times k \times h \times h! \quad (5.47)$$

The first factor $([p - (h + k)])$ is the number of remaining elements (it is the cardinality of the set Rem-Set). k is the number of active windows and $h \times h!$ is the number of comparisons between one element of Rem-Set and the elements of a given h window, by taking into account its permutational variation.

We leave it to the interested reader to determine the total number of pairwise comparisons from Eqs. 5.46 and 5.47.

5.4 Applying on σ Forms and Real Data

5.4.1 Preamble

Here, the objective is to realize the behavior of the different algorithms described in Sects. 5.3.1 and 5.3.2 in the case of first, simulated mathematical σ forms and second, on real data. Seven algorithms will be experienced. Their respective acronyms are

$$\mathbf{OP{-}LA,\ OP{-}SL,\ OP2{-}S,}$$
$$\mathbf{OP3{-}OSA,\ OP3{-}SSA,\ OP3{-}FIA,}\ \text{and}\ \mathbf{OP3{-}SAA.} \tag{5.48}$$

The general strategy is defined as follows. For a given algorithm among those listed in (5.48) and a given incidence data table, supposed to recover a seriation, we consider the series of seriation results obtained by applying the algorithm concerned on the data table. Attached with a given result, the respective values of the different criteria are examined.

These values are more or less, strongly related. However different profiles may sustain the values of these criteria. In these conditions, we retain either one, two or even three criteria in order to select the most efficient seriations.

Consider the case of a single criterion that we denote it by $Crit$ and suppose it has to be minimized. The seriations for which $Crit$ is minimal are selected, represented and examined relatively to the other criteria. Generally a single or a very few results are obtained.

Now, consider an ordered pair of selection criteria and let us denote it by $(Crit1, Crit2)$. Imagine we have to minimize $Crit1$ and to maximize $Crit2$. Denote by the following (5.49) the series of the obtained seriations by the algorithm concerned.

$$\{Ser(j)|1 \leq j \leq M\} \tag{5.49}$$

Suppose we are at the step j for which we have to evaluate $Ser(j)$ and designate by Ser_{opt} the best seriation already obtained strictly before the latter step. Then, denote by $(Crit1(opt), Crit2(opt))$ the value of the ordered pair of criteria $(Crit1, Crit2)$ on Ser_{opt}. In these conditions, if

$$Crit1(opt) > Crit1(Ser(j)) \text{ and } Crit2(opt) < Crit1(Ser(j))$$

substitute $Ser(j)$ to $Ser(opt)$

Thereby, the preference total order is the product of two total preorders associated with minimizing $Crit1$ and maximizing $Crit2$, respectively.

For simulated data, taking into account the mathematical definition and construction of these (see Sect. 5.4.2), it will be sufficient to use a single criterion to determine the most efficient seriations. As a matter of fact, it turns out that an optimum value of the criterion chosen, implies practically, optimal or near optimal values of the other criteria.

Real data (see Sect. 5.4.3) is much more complex. In this case, two criteria $Crit1$ and $Crit2$ have been used (see just above) in order to select the most efficient seriations. For this purpose, the criteria chosen are easy to design, intuitively speaking, and very different from each other.

The most often, ordering is first, carried out on the column attributes of the incidence data table. Ranking the row objects is then derived from ROFCO algorithm (see Sect. 5.2.3).

5.4.2   Simulated σ Forms

5.4.2.1   Introduction

These σ forms have been considered in Chap. 4 (see Figs. 4.1 and 4.2) and (see Sects. 4.4.2.2 to 4.4.2.6). The order in which their treatment will be described is the same as that of Chap. 4. The algorithms concerned are those listed in (5.48).

A given simulated σ form will be specified by its *type* defined by the following parameters:

- Number of blocks;

- For each of the ordered sequence of the different blocks (there are one, two or three): the row number, the column number, number of components equal to one per column.

Thus, the respective parameters of the σ forms, from left to right in Fig. 4.1 of Chap. 4, are:

$$[1, (21, 11, 11)], [1, (25, 16, 10)] \text{ and } [1, (25, 21, 5)] \tag{5.50}$$

Moreover, the respective parameters of the ordered sequence of the σ forms from left to right as given in Fig. 4.2 are

$$[2, (21, 11, 11), (21, 11, 11)] \tag{5.51}$$

$$\text{and}$$

$$[3, (15, 8, 8), (15, 8, 8), (15, 8, 8)]$$

It will be noted that in the case where the σ form includes more than a single block, the parameters of each of the blocks determining a σ form, are defined, relatively to the incidence data table by the rectangular part of it which circumscribes the block concerned.

The algorithms indicated in (5.48) and the configurations given in (5.50) and (5.51) give rise to 36 results. Each of them can be labeled by an ordered pair such as

$$(Algorithm, Configuration)$$

Thus, for example, for *OP2-S* algorithm, the first result (resp., the fourth one) will be labeled by the ordered pair

$$(OP2-S, [1, (21, 11, 11)])\qquad(5.52)$$

$$\text{resp.,}$$
$$(OP2-S, [2, (21, 11, 11), (21, 11, 11)])$$

Now, let us associate with each configuration a numerical table, whose rows representing the criteria and whose columns, the algorithms. In this, it is assumed that the numerical value in the cell at the intersection of a given row and a given column is defined by the measurement of the criterion on the algorithm concerned. In these conditions, a given result of the thirty six mentioned corresponds to a column of such a table. In this column, for the configuration, the respective values of the different criteria (see the end of Sect. 5.1) on the algorithm indexing the column are given.

Let us specify that before applying a given algorithm on a given configuration, a pair of independent random permutations on the rows and on the columns of the incidence data table are performed. This, even it can be shown that the result of the application of the algorithm is invariant whatever the pair of permutations.

A single selection criterion *Crit* is chosen in order to determine the most efficient seriations. The latter criterion—easy to visualize—is *Col-Row-Laz* (see Sect. 3.3.4.1, Chap. 3). Often, for most algorithms listed in (5.48), the *Crit* criterion value is null for the seriations selected. This is due to the formal definition and construction of the simulated seriations. We associate with each of the algorithms the optimal seriations obtained according to the selection criterion *Crit*. To each element of the latter seriations we attach the table of the values of the different criteria (*Lerm-Crit*, *Col-Laz*, *Row-Laz*, *Col-Row-Laz*, *ME4*, *ME8*, *Corr-8*, *DM-Corr*, *Moor-Strs*, and *Neum-Strs*). The quality of the results is remarkable. It shows all the power of combinatorial methods.

For results concerning a single block, we consider equally two σ forms configurations of the data table such as one of them is deduced from the other by respective inverse permutations of the column order and the row order. This is, for example, the case of the σ form moderately chained (see Sect. 5.4.2.2) for which both versions are figured.

We have limited ourselves to showing one version for the strongly chained and weakly chained σ forms (see Sect. 5.4.2.2).

In Sect. 5.4.2.3 σ forms comprising two or three blocks strongly chained are treated. For simplicity reasons, in each of both processings (for two blocks and three blocks), a unique result is given corresponding to different algorithms among those listed in (5.48). This unique result may represent several equivalent results. Recall that two results are equivalent if the first one can be deduced from the second by permuting the blocks and by permuting rows and columns inside the blocks, keeping the overall σ in two or three blocks.

5.4.2.2   Simulated σ Forms with One Block

Moderately Chained Case

In accordance with the above introductory section (see Sect. 5.4.2.1) the type of this σ form is [1, (25, 16, 10)]: One block, 25 rows, 16 columns and 10 one value per column. All of the algorithms described in this chapter and listed in (5.48) give the same perfect result associated with a null value of *Col-Row-Laz* criterion. Denoting by OP^* any of these algorithms, we obtain Fig. 5.1.

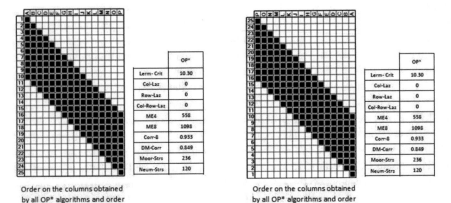

	OP*
Lerm- Crit	10.30
Col-Laz	0
Row-Laz	0
Col-Row-Laz	0
ME4	558
ME8	1098
Corr-8	0.933
DM-Corr	0.849
Moor-Strs	236
Neum-Strs	120

Order on the columns obtained
by all OP* algorithms and order
on the rows driven by the order
on the columns

	OP*
Lerm- Crit	10.30
Col-Laz	0
Row-Laz	0
Col-Row-Laz	0
ME4	558
ME8	1098
Corr-8	0.933
DM-Corr	0.849
Moor-Strs	236
Neum-Strs	120

Order on the columns obtained
by all OP* algorithms and order
on the rows driven by the order
on the columns

Fig. 5.1 Applying OP^* on moderately chained σ form

Fig. 5.2 Applying OP^* on strongly chained σ form

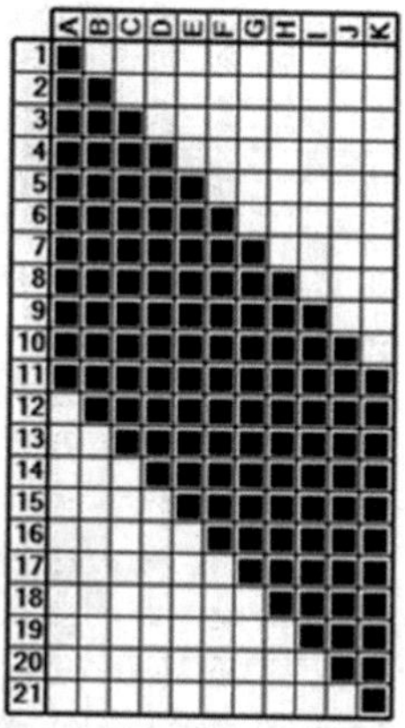

	OP*
Lerm- Crit	7.10
Col-Laz	0
Row-Laz	0
Col-Row-Laz	0
ME4	420
ME8	820
Corr-8	0.913
DM-Corr	0.707
Moor-Strs	156
Neum-Strs	80

Order on the columns obtained
by all OP* algorithms and order
on the rows driven by the order
on the columns

Strongly Chained Case

The type of this σ form is [1, (21, 11, 11)]: One block, 21 rows, 11 columns and 11 one value per column. All of the algorithms described in this chapter and listed in (5.48) give the same perfect result associated with a null value of *Col-Row-Laz* criterion. Denoting by OP^* any of these algorithms, we obtain Fig. 5.2.

Oddly,—compared with the above moderately chained case—the values of the criteria *Lerm-Crit*, $ME4$, $ME8$, *Corr-8*, *DM-Corr*, *Moor-Strs* and *Neum-Strs*, are less optimized.

Weakly Chained Case

The type of this σ form is [1, (25, 21, 5)]: One block, 25 rows, 21 columns and 5 one value per column. All of the algorithms described in this chapter and listed in (5.48) give the same perfect result associated with a null value of *Col-Row-Laz* criterion. Denoting by OP^* any of these algorithms, we obtain Fig. 5.3.

Now, let us compare the criteria values obtained in this case with those obtained in the case of strongly chained σ form. *Lerm-Crit*, *Corr-8* and specially *DM-Corr* are more optimized. On the other hand, $ME4$, $ME8$, *Moor-Strs* and *Neum-Strs* are less optimized. Everything happens as if the seriation phenomenon is more accentuated in the latter weak chained case.

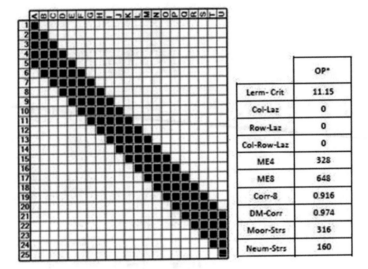

	OP*
Lerm- Crit	11.15
Col-Laz	0
Row-Laz	0
Col-Row-Laz	0
ME4	328
ME8	648
Corr-8	0.916
DM-Corr	0.974
Moor-Strs	316
Neum-Strs	160

Order on the columns obtained
by all OP* algorithms and order
on the rows driven by the order
on the columns

Fig. 5.3 Applying OP^* on weakly chained σ form

5.4.2.3   Simulated σ Forms with Two or Three Blocks

Two Blocks

The type of this σ form is [2, (21, 11, 11), (21, 11, 11)]: two equal blocks strongly chained and comprising each 21 rows, 11 columns and 11 one value per column.

Any OP^* algorithm brings out very clearly two blocks, without however providing exactly the same result. The two blocks can be swapped and respective inversions on rows and columns can occur.

There is a small difference in the overall shape of the result depending on the algorithm used. In the case of the algorithms $OP\text{-}LA$, $OP\text{-}SL$, $OP3\text{-}OSA$, $OP3\text{-}SSA$ and $OP3\text{-}SAA$, we obtain Fig. 5.4.

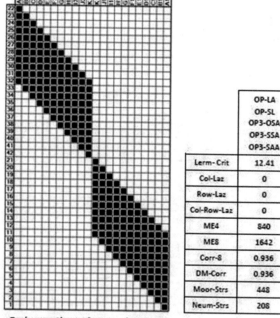

	OP-LA
	OP-SL
	OP3-OSA
	OP3-SSA
	OP3-SAA
Lerm-Crit	12.41
Col-Laz	0
Row-Laz	0
Col-Row-Laz	0
ME4	840
ME8	1642
Corr-8	0.936
DM-Corr	0.936
Moor-Strs	448
Neum-Strs	208

Order on the columns obtained
by OP-LA, OP-SL, OP3-OSA, OP3-SSA
and OP3-SAA algorithms and order
on the rows driven by the order on
the columns

Fig. 5.4 Applying *OP-LA, OP-SL, OP3-OSA, OP3-SSA* or *OP3-SAA* on two blocks strongly chained

In the case of *OP2-S* and *OP3-FIA* algorithms we obtain Fig. 5.5.

Now, let us compare the optimization criteria values with the case of moderately chained σ form [1, (25, 16, 10)]. It turns out that all of the criteria are more optimized with the two blocks σ structure [2, (21, 11, 11), (21, 11, 11)], except *Moor-Strs* and *Neum-Strs*. This behavior has to be interpreted in relation with seriation and clustering effects.

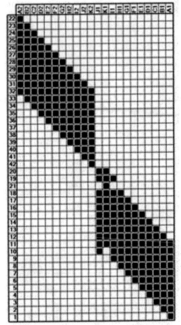

	OP2-S OP3-FIA
Lerm- Crit	12.38
Col-Laz	2
Row-Laz	1
Col-Row-Laz	3
ME4	834
ME8	1628
Corr-8	0.928
DM-Corr	0.935
Moor-Strs	476
Neum-Strs	220

Order on the columns obtained
by OP2-S and OP3-FIA algorithms
and order on the rows driven by
the order on the columns

Fig. 5.5 Applying *OP2-S*, or *OP3-FIA* on two blocks strongly chained

Three Blocks

The structure of this sigma form as represented in the introduction is [3, (15, 8, 8), (15, 8, 8), (15, 8, 8)]. Any OP * algorithm leads to an identical result in terms of the tabular representation of the seriation. The different results are obtained up to a permutation of the blocks and, for the same block, up to symmetrical joint inversions of the rows and the columns. Figure 5.6.

As previously, let us consider the comparison of the values of the optimized criteria with those obtained in the case of a single block given by the moderately chained sigma form. The same conclusions must be considered, that is, all criteria are more optimized with the three-block structure [3, (15, 8, 8), (15, 8, 8), (15, 8, 8)], except *Moor* − *Strs* and *Neum* − *Strs*.

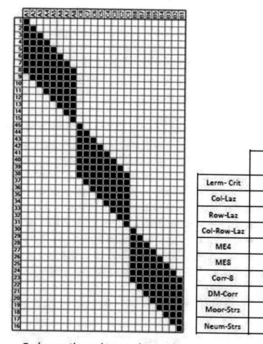

	OP*
Lerm- Crit	11.72
Col-Laz	0
Row-Laz	0
Col-Row-Laz	0
ME4	630
ME8	1222
Corr-8	0.917
DM-Corr	0.972
Moor-Strs	524
Neum-Strs	240

Order on the columns obtained
by OP-LA, OP3-FIA and OP3-SAA
algorithms and order on the rows
driven by the order on the columns

Fig. 5.6 Applying OP^*, on three blocks strongly chained

5.4.3 Real Data

5.4.3.1 Merovingian Belt Buckles-Plates

The two real data studied in Chap. 4 are reconsidered here. It is the "Merovingian buckles plates" (see 4.4.3.1) and the "La Tène Cemetery at Münsingen-Rain" (see Sect. 4.4.3.2).

Literature devoted to Merovingian Belt Buckles-Plates is impressive [4–6]. References concerning our own works are [2, 7], Chap. 18 of [8]. These are the works that are reconsidered here, but with new methods that are much richer, more analytical and more powerful.

As for simulated data, for a given algorithm, listed in (5.48), a method of selecting the most efficient variants is envisaged. The evaluation of the performances concerned is first visually evaluated in relation to the joint orderings of the columns and the rows. On the other hand, the respective values of the different criteria are examined. Precisely, relative to the seriations produced by a given algorithm, the optimization of a single criterion or more than one is used for the selection of the most efficient seriations.

In the simulated cases, the only criterion which intervened for the selection is *Col-Row-Laz*. A minimum value of this criterion was accompanied—in the case of a single block—by sufficiently optimized values of the other criteria. However, in the case of two or three blocks, the behavior of criteria $ME4$ and $ME8$ went in the opposite direction of optimization. In these conditions, we will use in this section—reserved for real data—after questioning the rule to adopt and in accordance with the description in the preamble (see Sect. 5.4.1), the product of two criteria $Crit1$ and $Crit2$ where $Crit1$ is defined by *Col-Row-Laz* and $Crit2$ by $ME8$.

As previously, a seriation algorithm is applied to order the set of the column attributes, the order on the row objects is then derived from ROFCO algorithm (see Sect. 5.2.3).

The same series of algorithms used as for the simulated data is employed.

OP-LA and **OP-SL** algorithms

Clearly, it is the phenomenon of seriation which dominates in this result (Fig. 5.7). The latter is more or less comparable—nevertheless slightly less efficient—with that obtained manually by Bertin (see Fig. 4.47 in Chap. 4).

OP2-S, **OP3-OSA**, **OP3-SSA** and **OP3-FIA** algorithms

All these four algorithms give the same result. This is a very good result showing a serial form. This result can be compared to that of the correspondence analysis (see Fig. 4.49 in Chap. 4). However, in this last result—as shown by the values of criteria $ME4$ and $ME8$—the phenomenon of clustering is better expressed. The shape obtained here is closer to Bertin's manual technique (see Fig. 4.47 in Chap. 4).

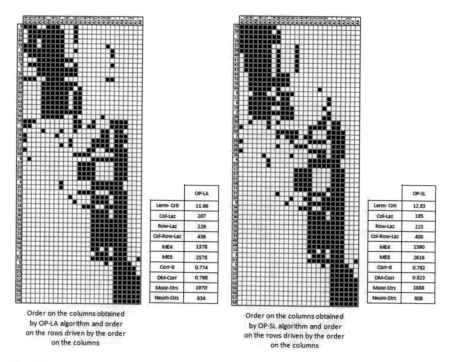

	OP-LA
Lerm- Crit	11.66
Col-Laz	207
Row-Laz	229
Col-Row-Laz	436
ME4	1378
ME8	2578
Corr-8	0.774
DM-Corr	0.798
Moor-Strs	1970
Neum-Strs	834

Order on the columns obtained
by OP-LA algorithm and order
on the rows driven by the order
on the columns

	OP-SL
Lerm- Crit	12.83
Col-Laz	185
Row-Laz	223
Col-Row-Laz	408
ME4	1390
ME8	2616
Corr-8	0.782
DM-Corr	0.823
Moor-Strs	1888
Neum-Strs	808

Order on the columns obtained
by OP-SL algorithm and order
on the rows driven by the order
on the columns

Fig. 5.7 Applying OP-LA and *OP-SL*, on Merovingian buckles-plates

OP3-SAA algorithm

In the case of using a single criterion for selecting a seriation, the iterative process assumed by the OP3-SAA algorithm is entirely linked to the optimization of this criterion. And this operates to the detriment of most of the other criteria. Indeed, this can be seen in the following three figures and the associated values of the criteria (see Figs. 5.9 and 5.10). In the first figure of Fig. 5.9 optimization of *Col-Row-Laz* is concerned and in the second one, it is ME8. Also, in the third figure (see Fig. 5.10) Moore-Strs criterion is considered. For each of the three cases, the criterion concerned is optimal. However and clearly, all of the others are not necessarily and this, even if a seriation structure turns out as relevant in the figure obtained (Fig. 5.8).

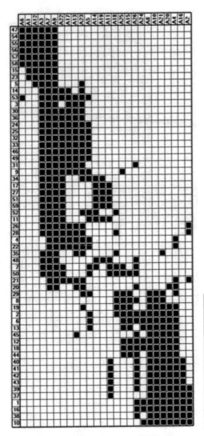

	OP2-S
	OP3-OSA
	OP3-SSA
	OP3-FIA
Lerm- Crit	13.20
Col-Laz	176
Row-Laz	215
Col-Row-Laz	391
ME4	1434
ME8	2704
Corr-8	0.814
DM-Corr	0.826
Moor-Strs	1706

Order on the columns obtained
by OP2-S, OP3-OSA, OP3-SSA and OP3-
FIA algorithms and order on the rows
driven by the order on the columns

Fig. 5.8 Applying *OP2-S, OP3-OSA, OP3-SSA* and *OP3-FIA* on Merovingian buckles-plates

It would be easy to examine all the possibilities, trying each time one of the criteria to be optimized. Our feeling is that none of the criteria can be on its own, the only criterion for selecting a seriation with regard to all the criteria.

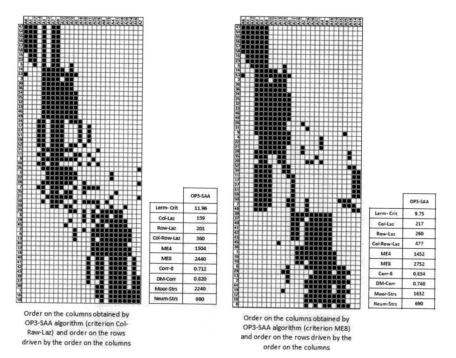

	OP3-SAA
Lerm- Crit	11.96
Col-Laz	159
Row-Laz	201
Col-Row-Laz	360
ME4	1304
ME8	2440
Corr-8	0.712
DM-Corr	0.820
Moor-Strs	2240
Neum-Strs	980

Order on the columns obtained by OP3-SAA algorithm (criterion Col-Raw-Laz) and order on the rows driven by the order on the columns

	OP3-SAA
Lerm- Crit	9.75
Col-Laz	217
Row-Laz	260
Col-Row-Laz	477
ME4	1452
ME8	2752
Corr-8	0.834
DM-Corr	0.740
Moor-Strs	1632
Neum-Strs	690

Order on the columns obtained by OP3-SAA algorithm (criterion ME8) and order on the rows driven by the order on the columns

Fig. 5.9 Applying *OP3-SAA* with a single criterion *Col-Row-Laz* or *M E*8

In these conditions and as announced in the preamble (see Sect. 5.4.1), selection of a seriation by optimization a single criterion cannot provide an effective solution on all the criteria. Then, we set about simultaneous optimization of different criteria whose respective behaviors tend to be opposed. For this, two criteria have been retained: *Col-Row-Laz* and *M E*8. The result obtained is excellent and better than previous ones, considering the compact shape of the serial form as well as the values of the various criteria, the latter being more or less optimal (see Fig. 5.11).

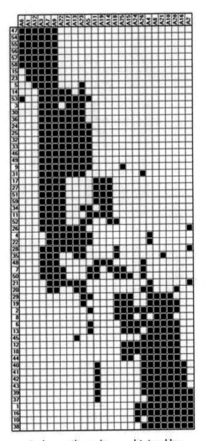

	OP3-SAA
Lerm- Crit	12.92
Col-Laz	177
Row-Laz	223
Col-Row-Laz	400
ME4	1446
ME8	2742
Corr-8	0.830
DM-Corr	0.829
Moor-Strs	1636
Neum-Strs	696

Order on the columns obtained by
OP3-SAA algorithm (criterion Moor-
Strs) and order on the rows driven by
the order on the columns

Fig. 5.10 Applying *OP3-SAA* with the single criterion *Moor-Strs*

Additionally, we have experienced optimization with a product of three criteria; namely, *Col-Row-Laz*, *ME8* and *Moor-Strs*. The result obtained is identical up to minor details. Above all, the criteria values are exactly the same as previously.

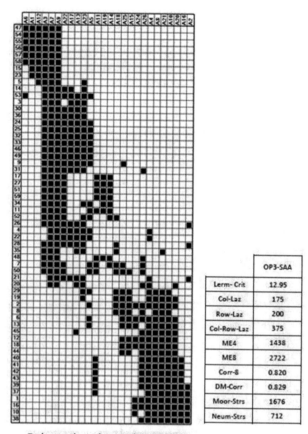

	OP3-SAA
Lerm-Crit	12.95
Col-Laz	175
Row-Laz	200
Col-Row-Laz	375
ME4	1438
ME8	2722
Corr-8	0.820
DM-Corr	0.829
Moor-Strs	1676
Neum-Strs	712

Order on the columns obtained by
OP3-SAA algorithm (criterion Col-Raw-Laz
and ME8) and order on the rows driven by
the order on the columns

Fig. 5.11 Applying OP3-SAA with the criterion product *Col-Row-Laz* and $ME8$

5.4.3.2 La Tène Cemetery of Münsingen-Rain

The following results are to be compared with that of Kendall, shown in the R software. Even, as has been noted, this latter result does not appear statistically natural. And this, because of a very large void (without one value) above a diagonal area.

In any case, as we have been able to express it, the various results obtained must be considered in a relative way. Between these results, the underlying methods and the expert's knowledge, a whole dialectic must be established.

OP-LA and **OP-SL** algorithms

These two algorithms give completely satisfactory and perfectly comparable results, clearly highlighting the serial form underlying these data. We are content to present the results of the first algorithm OP-LA (Fig. 5.12).

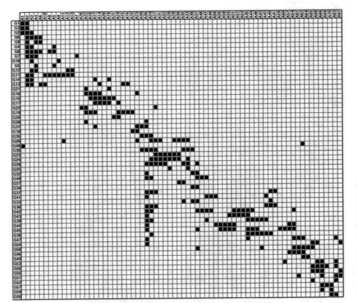

	OP-LA
Lerm- Crit	19.31
Col-Laz	281
Row-Laz	545
Col-Row-Laz	826
ME4	486
ME8	812
Corr-8	0.546
DM-Corr	0.929
Moor-Strs	2680
Neum-Strs	1190

Order on the columns obtained by OP-LA algorithm and
order on the rows driven by the order on the columns

Fig. 5.12 Applying *OP-LA* on Münsingen-Rain data

OP2-S OP3-OSA and **OP3-SSA** algorithms

Also, these three algorithms clearly highlight a form of seriation, accompanied by some relevant criteria. However, the order of the columns and rows can be quite different from Kendall's result (see Fig. 4.57). Let us give here one of these three results, namely, that of the OP2-S algorithm (Fig. 5.13).

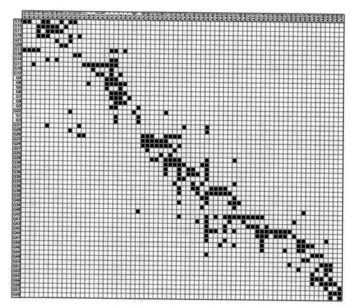

	OP2-S
Lerm- Crit	19.90
Col-Laz	291
Row-Laz	486
Col-Row-Laz	697
ME4	452
ME8	818
Corr-8	0.539
DM-Corr	0.951
Moor-Strs	2652
Neum-Strs	1252

Order on the columns obtained by OP2-S algorithm and
order on the rows driven by the order on the columns

Fig. 5.13 Applying *OP2-S* on Münsingen-Rain data

OP3-FIA and OP3-SAA algorithms

Recall that in the last two algorithms that we consider *OP3-FIA* and *OP3-SAA*, the junction of an element to a seriation under construction depends on the variation of a single arrangement in *OP3-FIA*, or on a sequence of adjacent arrangements in *OP3-SAA* (see Sects. 5.3.2.2 and 5.3.2.2).

With the *OP3-FIA* algorithm, the respective original rankings of the rows and columns are recovered. In addition, optimal results are obtained for the values of several criteria.

In the case *OP3-SAA* algorithm, the computational complexity tends to increase considerably. However, with this algorithm, one of the best of our results has surely been reached, including the comparison with the result of Kendall (see Fig. Ref Fig: Ord (Muns) data). Recall that in this algorithm, as in the previous ones, a simultaneous optimization, comprising $Col-Row-Laz$ and $ME8$, was applied in order to select the best seriations (Figs. 5.14 and 5.15).

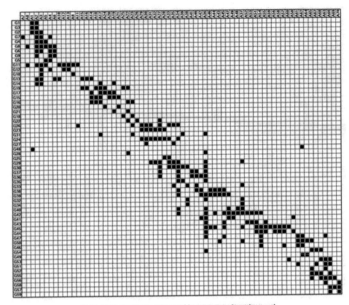

	OP3-FIA
Lerm- Crit	20.86
Col-Laz	258
Row-Laz	386
Col-Row-Laz	644
ME4	466
ME8	830
Corr-8	0.546
DM-Corr	0.959
Moor-Strs	2662
Neum-Strs	1236

Order on the columns obtained by OP3-FIA algorithm and
order on the rows driven by the order on the columns

Fig. 5.14 Applying *OP3-FIA* on Münsingen-Rain data

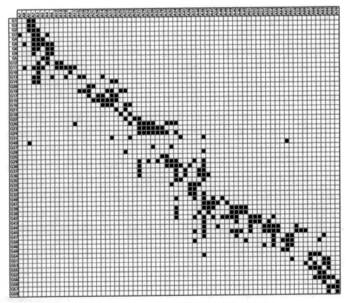

	OP3-SAA
Lerm- Crit	20.38
Col-Laz	241
Row-Laz	348
Col-Row-Laz	589
ME4	482
ME8	904
Corr-8	0.589
DM-Corr	0.958
Moor-Strs	2514
Neum-Strs	1204

Order on the columns obtained by OP3-SAA algorithm and
order on the rows driven by the order on the columns

Fig. 5.15 Applying *OP3-SAA* on Münsingen-Rain data

References

1. M. Hahsler, K. Hornik, C. Buchta, Getting things in order: an introduction to the R package seriation. J. Stat. Softw. **25**(3), 1–34 (2008)
2. H. Leredde, *La méthode des poles d'attraction, La méthode des poles d'agrégation*. Ph.D. thesis, Université de Paris 6, Oct. 1979
3. I.C. Lerman, *Foundations and Methods in Combinatorial and Statistical Data Analysis and Clustering* (Springer, 2016)
4. B. Effros, I. Moreira, *The Oxford Handbook of the Merovingian World*, ed. by B. Effros, I. Moreira (Oxford University Press, 2020)
5. P. Perin, La datation des tombes mérovingiennes. historiques, méthodes, applications. *Paris, DROZ* (1980)
6. H.G. Semanville, Les plaques-boucles mérovingiennes ornées d'une croix encadrée par deux griffons: à propos d'une découverte faite à Fleurey-sur-Ouche. Revue Archeologique de l'Est **59**(2), 585–602 (2010)
7. H. Leredde, P. Perin, Les plaques boucles mérovingiennes. Dossiers de l'Archéologie **42**, 83–87, Mars-Avril 1980
8. I.C. Lerman, *Classification et analyse ordinale des données*. Dunod and http://www.brclasssoc. org.uk/books/index.html (1981)

Chapter 6
Clustering Methods from Proximity Variance Analysis

6.1 Introduction and General Presentation

Two new families of clustering algorithms emerge from the a priori determination of attraction poles.

Denote by **E** the set to be clustered. The first family operates by successive divisions of **E**, around attraction poles. For a given algorithm of this family the first division (separation) is into two classes, around, respectively, the two first poles. In the progression of the algorithm, each element, not yet classified is assigned to one or, exclusively, to the other of the two poles. The $(k-1)$th separation (division) of the whole set is into k classes around, respectively, the first k attraction poles. In the assignation process, each element not yet classified is assigned to exactly one of the k poles.

With a given algorithm of this family, a decendant but non-hierarchical sequence of partitions (classifications) is carried out. We call this family *APMSR* (Attraction Pole Methods operating by Successive Reallocations).

For an algorithm belonging to the second family called Attraction Pole Methods operating by Successive Aggregations (*APMSA*), a partition of the entire set is built, in a recurrent way, class after class. The first class is trained around the single (the first one) attraction pole defined at the level of the entire set **E**. In the algorithm progression, a given class is trained around the first attraction pole defined relatively to the **E** subset not yet decomposed into classes.

We assume the set **E** provided with a similarity, dissimilarity or a distance function. The cardinality of **E** is denoted here by m.

As just mentioned, in the first step of an algorithm of the second family, the attraction pole driving the first class is established with respect to the entire set **E**. Suppose, to fix ideas, that **E** is provided with a distance measure d and let P designate this attraction pole. Now, consider the distance distribution to P, denoted as follows:

$$\{d(P, x) | x \in \mathbf{E} - \{P\}\} \tag{6.1}$$

Associate with it the increasing sequence

$$d(P, x_{(1)}) \leq d(P, x_{(2)}) \leq \cdots \leq d(P, x_{(j)}) \leq \cdots \leq d(P, x_{(m-1)}) \tag{6.2}$$

In these conditions, the first class can be obtained by grouping all of the elements of the sub-sequence

$$\{x_{(j)} | 0 \leq j \leq l\} \tag{6.3}$$

where $x_{(0)} = P$ and $l < m$.

The boundary element $x_{(l)}$ can be determined by using a criterion enabling to situate $d(P, x_{(l+1)})$ with respect to $d(P, x_{(l)})$. This criterion has to be specified adequately (see Sect. 6.3).

Now, let us examine the principle governing the mutual positions of the respective attraction poles. To fix ideas without restricting generality, we can consider the simple case of two attraction poles. Suppose already determined the first of them that we designate by P_1.

For the determination of the second pole P_2, we begin by recalling the case of planar geometrical representation around two attraction poles (see Sect. 2.4.5, Chap. 2). In the latter, it is the principle of maximal *neutrality* which governs the determination of P_2 with respect to P_1. More explicitly, according to the notations adopted in Sect. 6.2.1, $S(P_1, P_2)$ (resp., $P(P_1, P_2)$) has to be the nearest null value.

We consider the context in which the data is an incidence data table and where the objective is clustering the set of column attributes or that, of the row objects. The latter clustering (of the row objects) can be get directly or deduced from that of the column attributes. The techniques developed are based on *similarity, dissimilarity* or *distance* notions. Therefore, they can be extended to any type of data table (see Chaps. 3–7 of [1]). In Sect. 6.2.1 we will specify the similarity, dissimilarity or distance indices (coefficients) between column attributes or row objects of an incidence data table. Let $\mathbf{E}$ designate the set concerned (column attributes or row objects). The techniques enabling us to detect in $\mathbf{E}$ a system of attraction poles will be defined in Sect. 6.2.2. Each of the attraction poles will drain off a cluster. In Sect. 6.2.3 we will express several association criteria comparing one element of $\mathbf{E}$, not yet assigned to an attraction pole, to an aggregate being trained around a given pole.

In this chapter we are concerned with *clustering*. There are two fundamental principles for the attraction poles determination and their associated clusters. The first one, just expressed, is of mutual maximal *independence*. The second one is that of mutual maximal *opposition*. For the independence (neutrality) principle, the basic comparison coefficient between attributes is $S(k, l)$ (see (6.33) in Sect. 6.2.1). And, in the case of object comparison, the coefficient is $P(h, i)$ (see (6.37) in Sect. 6.2.1).

Relative to the opposition principle in building clusterings by the pole attraction method, the basic distance measure for comparing attributes is given by $D_\alpha^2(k, l)$ and that, for comparing objects, by $D_\omega^2(k, l)$ (see (6.38) and (6.39) in Sect. 6.2.1).

According to the notations adopted in Sect. 2.4.5 of Chap. 2, for a given x element of $\mathbf{E}$, $\mathcal{V}(x)$ (resp., $\mathcal{M}_2(x)$) will denote the variance (resp., the absolute moment of order 2) of the distribution of the distances of the different elements of $\mathbf{E}$ to x.

Thereby, two options may be envisaged for defining a criterion which enables attraction poles to be determined. The first one is based on $\mathcal{V}(x)$ and the second, on $\mathcal{M}_2(x)$. One of the general form of the latter criterion when $\mathcal{V}(x)$ is employed is given in (2.108) of Chap. 2. More general statements will be given in Sect. 6.2.2.

In Sect. 6.2.3 we will detail the different techniques for forming a classification (clustering) from the determination of a collection of attraction poles. Therefore, the mentioned *APMSR* methods will be made explicit.

A fundamental matter for these methods consists of managing the descendant, but non-hierarchical sequence of classifications obtained. More precisely, we have to retain among these clusterings, some of them in accordance with the user need.

Two criteria will contribute to the latter purpose. The first one is the classical within (explained) inertia (Ward criterion) and the second one is that employed in order to detect the *significant* levels and *significant* nodes of a classification tree in hierarchical clustering (see Sects. 9.3.4–9.3.6 of [1]). The use of these criteria—for retaining some interesting classifications from the clustering sequence carried out by a pole attraction method—will be expressed in Sect. 6.4.

In Sect. 6.3 we will specify the second algorithm family that we have—just above—entitled *APMSA*.

In the section devoted to the developments (see Sect. 6.5), we will point out an original extension of the latter type of methods (*APMSA*) for hierarchical clustering algorithms. Also and mainly, we will emphasize the role of the divisive *APMSR* methods (the first method family) in the K-means algorithms [2, 3].

The application of the new clustering methods in real or simulated cases is discussed in the last section (see Sect. 6.6)

6.2 The Method Family of Attraction Poles by Successive Reallocations (*APMSR*)

6.2.1 Data, Similarities and Distances

The data is defined by a finite set $\mathbf{E}$ endowed with a similarity, dissimilarity or distance function. In practice, the index is normalized by considering the nature and the structure of the set $\mathbf{E}$ to be clustered. We will clarify this normalization in the cases of interest for which the set to be classified is that of the column attributes or that, of the row objects of an incidence (presence-absence) data table.

Take up again here (2.1) and (2.2) from Chap. 2:

$$\mathcal{O} = \{o_i | 1 \le i \le n\}$$

and

$$\mathcal{A} = \{a^j | 1 \le j \le p\} \tag{6.4}$$

where $\mathcal{O}$ and $\mathcal{A}$ are the set of the objects and that of the descriptive Boolean attributes, respectively.

The incidence data table denoted by $\mathcal{T}$ can be written as follows:

$$\mathcal{T} = \{x_i^j | 1 \le i \le n, 1 \le j \le p\} \tag{6.5}$$

where $(\forall (i, j), 1 \le i \le n, 1 \le j \le p)$, $x_i^j = 1$(resp., 0) if the Boolean attribute a^j is *THRUE* (resp., *FALSE*) in the object o_i, $1 \le i \le n, 1 \le j \le p$.

The reference point of the construction of the distance indices between column attributes and between row objects is defined by the association coefficient $S(k, l)$ (see 2.9) of Chap. 2. Recall its expression:

$$S(k, l) = \frac{s(k, l) - n.\eta_k.\eta_l}{\sqrt{n.\eta_k.\eta_l}} \tag{6.6}$$

where

$$s(k, l) = \sum_{1 \le i \le n} x_i^k . x_i^l \tag{6.7}$$

and where η_k (resp., η_l) is the proportion of components equal to 1 in the kth (resp., lth) column of the incidence data table $\mathcal{T}$.

The data table $\mathcal{T}'$ is obtained from $\mathcal{T}$ by substituting the measures $x_i^{\prime j}$ for the measures x_i^j:

$$x_i^{\prime j} = \frac{x_i^j - \eta_j}{\sqrt{\eta_j . \sqrt{n}}} \tag{6.8}$$

$1 \le i \le n, 1 \le j \le p$

Therefore, as already expressed in (2.17) of Chap. 2, we have

$$S(k, l) = \sum_{1 \le i \le n} x_i^{\prime k} . x_i^{\prime l} \tag{6.9}$$

$1 \le k \le l \le p$

The associated similarity index between row objects can be written as follows:

$$P(h, i) = \sum_{1 \le j \le p} x_h^{\prime j} . x_i^{\prime j} \tag{6.10}$$

In these conditions, the following two distance functions are set up. The first one denoted by D_α is between column attributes and the second one D_ω is between row objects:

$$D_\alpha^2(k, l) = \sum_{1 \le i \le n} (x_i'^k - x_i'^l)^2 \tag{6.11}$$

$1 \le k \le l \le p$ and

$$D_\omega^2(h, i) = \sum_{1 \le j \le p} (x_h'^j - x_i'^j)^2 \tag{6.12}$$

$1 \le h \le i \le n$.

This strategy is of the same nature as that applied in Principal Component Analysis where the distances between variables on one hand and between objects on the other one are calculated directly from the table of normalized (centered and reduced) measures, with respect to the respective statistical distributions of the descriptive variables [4–6].

According to the case concerned (clustering the attribute set or clustering the object set) the index D_α or that D_ω can play the role of D in Sect. 3.4.

It is of importance to recall here the relation which connects $S(k, l)$ to $D_\alpha^2(k, l)$ (see 2.89):

$$S(k, l) = \frac{1}{2}\{\sqrt{n}.[(1 - \eta_k) + (1 - \eta_l)] - D_\alpha^2(k, l)\} \tag{6.13}$$

which results from

$$S(j, j) = \sqrt{n}.(1 - \eta_j) \tag{6.14}$$

$1 \le j \le p$

An equivalent formulation to (6.13) is given by

$$D_\alpha^2(k, l) = S(k, k) + S(l, l) - 2S(k, l) \tag{6.15}$$

In a dual way, we also have to consider the relation which connects $P(h, i)$ to $D_\omega^2(h, i)$. We have

$$P(i, i) = \sum_{1 \le j \le p} x_i'^{j2} = \sum_{1 \le j \le p} \frac{(x_i^j)^2 - 2.\eta_j x_i^j + \eta_j^2}{\eta_j.\sqrt{n}} \tag{6.16}$$

Now, introduce for the ith row the proportion γ_i of components equal to one, $1 \le i \le n$. We have

$$P(i, i) = \frac{1}{\sqrt{n}}.[\sum_{1 \le j \le p} \frac{x_i^j}{\eta_j} - 2.p.\gamma_i + \sum_{1 \le j \le p} \eta_j] \tag{6.17}$$

In these conditions, we have

$$D_\omega^2(h, i) = P(h, h) - 2.P(h, i) + P(i, i) \tag{6.18}$$

And then, we obtain

$$P(h, i) = \frac{1}{2}.[P(h, h) + P(i, i) - D_\omega^2(h, i)] \tag{6.19}$$

6.2.2 Determination of a System of Attraction Poles

As pointed out in the introduction section (see Sect. 6.1), it is the maximal mutual dissimilarity principle which prevails. We consider two main versions of applying this principle. For the first one, the dissimilarity concerned is relative to *independence* or *neutrality* and for the second, to *opposition*. In the following, d will designate the dissimilarity concerned.

Now, let us consider the euclidean representation defined in (6.8). For this, the Boolean attribute a^j is represented by the vector

$$x^j = {}^T(x_1'^j, x_2'^j, \ldots, x_i'^j, \ldots, x_n'^j) \tag{6.20}$$

and the object o_i, by the vector

$$x_i = {}^T(x_i'^1, x_i'^2, \ldots, x_i'^j, \ldots, x_i'^p) \tag{6.21}$$

where the upscript T indicates *Transpose*, $1 \le i \le n, 1 \le j \le p$.

In fact, in the clustering attraction pole methodology, it is the square of the dissimilarity d which is directly used. For this, let us consider four main cases:

1. Attribute clustering and independence dissimilarity, for this case

$$d^2 = \frac{1}{S^2(k, l)} \tag{6.22}$$

$1 \le k < l \le p$ (see (6.9));
2. Attribute clustering and opposition dissimilarity, for this case

$$d^2 = D_\alpha^2 \tag{6.23}$$

$1 \le k < l \le p$ (see (6.11));
3. Object clustering and independence dissimilarity, for this case

$$d^2 = \frac{1}{P^2(h, i)} \tag{6.24}$$

$1 \le h < i \le n$ (see (6.10));

4. Object clustering and opposition dissimilarity, for this case

$$d^2 = D_\omega^2(h, i) \qquad (6.25)$$

$1 \le h < i \le n$ (see (6.12)).

Notice here that $S(k, l)$ (resp., $P(h, i)$) is nothing else than the inner scalar product between two vectors of the form (6.9) (resp., of the form (6.10)). On the other hand, recall the equations (6.15), (6.17)–(6.19), relating the independence to opposition dissimilarities.

To begin, we attach with a given element x of the set $\mathbf{E}$, a numerical index translating the distance variability to x. As mentioned in the introduction above (see Sect. 6.1), two coefficients can be considered: the variance $V(x)$ and the absolute moment of order 2, $\mathcal{M}_2(x)$. We have

$$V(x) = \frac{1}{m-1} \sum \{(d(x, y) - \bar{d}(x))^2 | y \in \mathbf{E} - \{x\}\}$$

where

$$\bar{d}(x) = \frac{1}{m-1} \sum \{d(x, y) | y \in \mathbf{E} - \{x\}\}$$

and

$$\mathcal{M}_2(x) = \frac{1}{m-1} \sum \{(d(x, y))^2 | y \in \mathbf{E} - \{x\}\} \qquad (6.26)$$

Clearly, we have

$$V(x) = \mathcal{M}_2(x) - (\bar{d}(x))^2 \qquad (6.27)$$

In the following $W(x)$ will indicate one of the both Statistics $V(x)$ or $\mathcal{M}_2(x)$. In the subsequent calculus, one of these two coefficients can be chosen.

Similarly to (2.94) of Chap. 2—but here we work with distances—the first pole P_1 is defined by

$$P_1 = Arg\{max(W(x)) | x \in \mathbf{E}\} \qquad (6.28)$$

Let us consider now the problem of fixing the second pole P_2 ($P_1 \ne P_2$). P_2 has to satisfy jointly the two following conditions:

1. $W(P_1)$ as large as possible;
2. $d(P_1, P_2)$ as large as possible.

Consequently, we propose the following criterion which combines the two preceding conditions

$$W(P_2) \times d^2(P_1, P_2) \qquad (6.29)$$

Thereby, conditions 1 and 2 take part analytically in the same manner and with equal importance.

In fact, the two multiplicative components of (6.29) are of the same analytical nature (dimension), the latter corresponding to a distance square.

In these conditions P_2 is defined as

$$P_2 = Arg\{max(W(x).d^2(P_1, x))|x \in E - \{P_1\}\} \tag{6.30}$$

More generally, for determining the pole sequence by using the maxmin criterion, the jth pole is obtained recursively from an equation comparable with (2.108) of Chap. 2, namely,

$$P_j = Arg\{max\{min(W(x).d^2(P_h, x))|1 \le h \le j - 1, x \in E - \{P_h|1 \le j - 1\}\}\} \tag{6.31}$$

$j = 3, 4, \ldots$

A different type of a *maxmin* criterion may equally be considered. For example, the following determination rule—based on product mean—of the jth pole P_j deserves to be studied—at least experimentally—in relation with (6.10).

$$P_j = Arg\{(\Pi_h W(x).d^2(P_h, x))^{\frac{1}{j-1}}|1 \le h \le j - 1, x \in E - \{P_h|1 \le j - 1\}\} \tag{6.32}$$

where Π is the product symbol, $j = 3, 4, \ldots$.

6.2.3   Cluster Formation Around the Attraction Poles

6.2.3.1   General Principle of the Algorithm

Suppose established a set $\mathcal{P}$ of K attraction poles. Denote this set as follows:

$$\mathcal{P} = \{P_j|1 \le j \le K\} \tag{6.33}$$

The initial partition of $\mathcal{P}$ is into K classes where each of them is a singleton class including one of the attraction poles (one element of $\mathcal{P}$). By denoting C_0 such a clustering, C_0 can be expressed as follows:

$$C_0 = \{\{P_j\}|1 \le j \le K\} \tag{6.34}$$

The objective consists of forming a clustering of the entire set E by assigning each of its elements to one of the K attraction poles. The algorithm progression consists of assigning a not yet classified element of E, to a cluster, in its process to be set up around one of the poles. Let us designate by C_l^j the cluster constituted around the jth pole P_j, after the assignment of the lth element of E, counting the K attraction poles, $1 \le j \le K$, $K + 1 \le l \le m - 1$. Thereby, the subset of E already classified comprises l elements where l equals K plus the number of elements assigned to the different poles. In these conditions, designate by R_l the remaining set of the elements to be classified. We have

$$R_l = \mathbf{E} - \bigcup_{1 \leq j \leq K} C_l^j$$

$$card(R_l) = m - l \tag{6.35}$$

In these conditions, the imputation of the $(l + 1)$th element x of $\mathbf{E}$ to the cluster constituted around one of the attraction poles, say P_k, will depend on a comparison criterion between x and C_l^k. Let us denote this criterion in the following form

$$Crit(C_l^k, x) \tag{6.36}$$

By supposing, without generality restriction, that (6.36) is defined by a similarity or dissimilarity function between C_l^k and x, x will be assigned to the kth class (cluster) around the attraction pole P_k:

$$C_l^k \leftarrow \{x\} \tag{6.37}$$

where the ordered pair (C_l^k, x) realizes

$$optimum\{Crit(C_l^k, x) | 1 \leq k \leq K, x \in R_l\} \tag{6.38}$$

This optimum is a maximum if it is defined by a similarity, and a minimum if the criterion is defined by a dissimilarity or distance.

In these conditions, a sorting is carried out in order to obtain this optimization. This sorting is over $(m - l) \times K$ arguments. Indeed, $card(R_l) = m - l$. And then, all the assignments concerned by the optimum value of $Crit(C_l^k, x)$ are performed.

6.2.4 Assignment Criteria

6.2.4.1 Preamble

We shall give here different versions of the assignment criterion $Crit(C_l^k, x)$ (see (6.36)) of the element x—not yet aggregated—to the cluster C_l^k already formed around the kth pole P_k. Two alternatives can be considered for defining each of the criterion version of $Crit(C_l^k, x)$. The first one is based on maximizing a similarity and the second one on minimizing a dissimilarity or distance, between the singleton class $\{x\}$ and C_l^k. We have established in Sect. 6.2.1 similarity and distance indices for comparing two single elements of the set to be clustered (the column attribute or the row object sets). We shall detail here seven different versions of the criterion $Crit(C_l^k, x)$ based on similarity and also their respective associated criteria based on distance. The first six criteria will be denoted as follows:

$$SimCrit_0(C_l^k, x) \,,\, SimCrit_{SL}(C_l^k, x) \,,\, SimCrit_{CL}(C_l^k, x)$$
$$SimCrit_{AL}(C_l^k, x) \,,\, SimCrit_{PAL}(C_l^k, x)$$
$$SimCrit_W(C_l^k, x) \,,\, SimCrit_{LLA}(C_l^k, x) \tag{6.39}$$

Besides, the respective associated criteria will be denoted

$$DisCrit_0(C_l^k, x) \,,\, DisCrit_{SL}(C_l^k, x) \,,\, DisCrit_{CL}(C_l^k, x)$$
$$DisCrit_{AL}(C_l^k, x) \,,\, DisCrit_{PAL}(C_l^k, x)$$
$$DisCrit_W(C_l^k, x) \,,\, DisCrit_{LLA}(C_l^k, x) \tag{6.40}$$

More explicitly, with notations used above,

1. (i) In the case of comparing column attributes, the criteria in (6.39) refer to the similarity coefficient $S(k, l)$;
2. (ii) In the case of comparing row objects, the criteria in (6.39) refer to the similarity index $P(h, i)$;
3. (iii) In the case of comparing column attributes, the criteria in (6.40) refer to the distance index $D_\alpha(k, l)$;
4. (iv) In the case of comparing row objects, the criteria in (6.40) refer to the distance index $D_\omega(h, i)$.

In fact, the subscripts *SL*, *CL*, *AL*, *PAL*, *W*, and *LLA* refer to comparison between two disjoint classes in ascendant hierarchical clustering: *SL* for Single Linkage, *CL* for Complete Linkage, *AL* for Average Linkage, *W* for Ward criterion (defined by the inertia variation) and *LLA* for the Likelihood Linkage Analysis criterion (see Sect. 10.3 of Chap. 10 of [1]).

Now, we shall detail the family of criteria (6.39) on the one hand and that of (6.40) on the other. Concerning (6.39) (**E**, *S*) will designate the ordered pair composed of the set **E** (the column attributes or the row objects) and the similarity index defined on **E**, according to 1 and 2 of the above items.

Relative to the development of (6.40), (**E**, *D*) will designate the ordered pair composed of the set **E** (the column attributes or the row objects) and the distance measure defined on **E**, according to 3 and 4 of the above items.

When (**E**, *S*) is involved, maximization procedures are considered and when it is the case of (**E**, *D*), minimization is involved.

Moreover, the first criterion in (6.39) (resp., (6.40)) is defined by the similarity (resp., distance) between x and the kth pole P_k

Now, we shall present in the framework of the *APMSR* family of methods, the different criteria listed in (6.39).

6.2.4.2   Criteria Listed in (6.39)

$$SimCrit_0(\mathbf{C_1^k}, \mathbf{x}) = S(x, P_k)$$

where

$$P_k = Arg\{max\{S(x, P_j)|1 \leq j \leq K\}\}$$

Similarly, in what follows, if a criterion corresponds to a similarity measure between x and C_l^k, it is necessary to determine the value of k which maximizes this measure. This, in order to pinpoint the cluster to which x has to be assigned.

$$SimCrit_{SL}(\mathbf{C_1^k}, \mathbf{x}) = max\{S(x, y)|y \in C_l^k\}$$

$$SimCrit_{CL}(\mathbf{C_1^k}, \mathbf{x}) = min\{S(x, y)|y \in C_l^k\}$$

$$SimCrit_{AL}(\mathbf{C_1^k}, \mathbf{x}) = \frac{1}{card(C_l^k)} \sum\{S(x, y)|y \in C_l^k\}$$

The preceding criterion is in the form of an equally weighted mean of similarities. The one $SimCrit_{PAL}(\mathbf{C_1^k}, \mathbf{x})$ (*P* for progressive) is a variant by introducing a weighting in the form (5.16). It will be used in our experiments.

$$SimCrit_W(\mathbf{C_1^k}, \mathbf{x}) = card(C_l^k) \times \frac{1}{S^2(x, G_l^k)}$$

where G_l^k is the center of gravity of the class in the process of formation, C_l^k:

$$G_l^k = \frac{1}{card(C_l^k)} \sum\{y|y \in C_l^k\} \tag{6.41}$$

This criterion is based on inertia. It is defined by the inertia moment of the cluster C_l^k with respect to x. C_l^k is represented by the ordered pair $(G_l^k, card(C_l^k))$ where G_l^k is the gravity center of C_l^k. The distance index concerned ($\frac{1}{S^2}$) brings out distance with respect neutrality or independence.

The calculation of the similarity S between x and G_l^k is performed from vector description of x and G_l^k. This, in the same way as we have to compare two vector Boolean descriptions (the same analytical form).

One option may consist of substituting for G_l^k the most central element O_l^k of C_l^k. O_l^k is defined as the element of C_l^k which minimizes the sum of its square distances to the other elements of C_l^k. This sum refers to an euclidean representation of **E**.

Let us now detail

$$SimCrit_{LLA}(C_l^k, x)$$

The first step to establish this criterion consists of substituting a probabilistic similarity index to the similarity coefficient S, defined on the set $\mathbf{E}$ to be clustered. For this, refer to the series of Eqs. (7.2.3)–(7.2.6) of Sect. 7.2.1 of Chap. 7 of [1]. According to (7.2.5), we start by defining a standardized similarity index as follows:

$$Q^g(y, z) = \left(\frac{S(y, z) - mean_e(S)}{\sqrt{var_e(S)}} \right) \tag{6.42}$$

where $mean_e$ and var_e designate, respectively, the empirical mean and variance of S, over the set $P_2(\mathbf{E})$ of unordered distinct element pairs of $\mathbf{E}$. To be more explicit, we have

$$mean_e(S) = \frac{1}{\binom{m}{2}} \sum \{S(y, z) | \{y, z\} \in P_2(\mathbf{E})\} \tag{6.43}$$

$$var_e(S) = mean_e(S^2) - \left(mean_e(S) \right)^2 \tag{6.44}$$

where $m = card(E)$ and where $\binom{m}{2}$ is a binomial coefficient.

In these conditions, the table of the similarity indices (7.2.6) of Chap. 7 of [1] has to be taken up again. Let us recall here its expression:

$$\mathcal{P}^g = \{P^g(y, z) = \Phi(Q^g(y, z)) | \{y, z\} \in P_2(\mathbf{E})\} \tag{6.45}$$

where Φ is the cumulative normal distribution function.

Therefore, as just mentioned above, by referring to Eqs. (10.4.7) and (10.4.8) of Sect. 10.4 of Chap. 10 of [1], for a given value ϵ of the LLA criterion ($0 \le \epsilon \le 1$), we can write

$$Crit_{LLA}(C_l^k, x) = -log_2\left(-log_2(LL_{max}^\epsilon(C_l^k, \{x\})) \right) \tag{6.46}$$

where

$$LL_{max}^\epsilon(C_l^k, \{x\}) = \left(p(C_l^k, \{x\}) \right)^{\pi(l,k)} \tag{6.47}$$

where $\pi(l, k) = (card(C_l^k))^\epsilon$ and where

$$p(C_l^k, \{x\}) = max\{P^g(y, x) | y \in C_l^k\} \tag{6.48}$$

A final, but very important point is the following. A partition into K clusters obtained by any of the *APMSR* methods permits the $K-means$ algorithm to be initialized efficiently. This will be emphasized once more in Sect. 6.5, reserved to the developments.

6.2.4.3   Criteria Listed in (6.40)

This list is somewhat parallel to the previous one. Despite some repetition, we want to produce it. Let us recall once again, that the dissimilarity underlying the previous list reflects *independence* or *neutrality*, whereas here, it translates *opposition* (see the preamble in Sect. 6.2.4.1). If it is the attribute clustering, D corresponds to D_α and if it is the object clustering D corresponds to D_ω (see the enumeration following (6.40)).

$$DisCrit_0(\mathbf{C}_1^k, \mathbf{x}) = \mathbf{D}(\mathbf{x}, \mathbf{P_k})$$

where

$$P_k = Arg\{min\{D(x, P_j)|1 \le j \le K\}\}$$

Similarly, in what follows, a criterion corresponding to a dissimilarity or distance measure between x and C_1^k, needs the determination of the value of k which minimizes the value of this measure. This in order to pinpoint the cluster to which x has to be assigned.

$$DisCrit_{SL}(\mathbf{C}_1^k, \mathbf{x}) = min\{D(x, y)|y \in C_1^k\}$$

This criterion is defined by the *minimum* distance D between the singleton class including the single element x and the class C_1^k, in the process formation.

$$DisCrit_{CL}(\mathbf{C}_1^k, \mathbf{x}) = max\{\mathbf{D}(\mathbf{x}, \mathbf{y})|\mathbf{y} \in \mathbf{C}_1^k\}$$

This criterion is associated with the *maximal* distance D between the singleton class including the element x and the class in the process of formation, C_1^k.

$$DisCrit_{AL}(\mathbf{C}_1^k, \mathbf{x}) = \frac{1}{\mathbf{card}(\mathbf{C}_1^k)} \sum \{\mathbf{D}(\mathbf{x}, \mathbf{y})|\mathbf{y} \in \mathbf{C}_1^k\}$$

This criterion is defined by the distance *average* between the singleton class including the element x and the class in the process of formation, C_1^k. *card* stands for the cardinal function.

Given that the criterion is in the form of an equally weighted mean of distances, the variant $DisCrit_{PAL}(\mathbf{C}_1^k, \mathbf{x})$ (*P* for progressive) is considered by introducing a weighting in the form (5.16). In this case the distance criterion is denoted *PAL* (*P* for progressive). It will be used in our experiments.

$$DisCrit_W(\mathbf{C}_1^k, \mathbf{x}) = \mathbf{card}(\mathbf{C}_1^k) \times \mathbf{D}^2(\mathbf{x}, \mathbf{G}_1^k)$$

According to (10.3.12) and (10.3.13) of Sect. 10.3.3 of Chap. 10 of [1], this criterion corresponds in agglomerative hierarchical clustering to the variation of the inertia associated with merging two classes. It is equal to the inertia moment of the centers of gravity of the two classes, these being weighted by their respective cardinalities.

Now, let us describe the criterion

$$DisCrit_{LLA}(C_l^k, \mathbf{x})$$

The following description is very similar to that of $SimCrit_{LLA}(C_l^k, \mathbf{x})$, considered above. We just have to manage distance instead of similarity. We start by defining a standardized similarity index as follows:

$$Q^g(y, z) = -\left(\frac{D(y, z) - mean_e(D)}{\sqrt{var_e(D)}}\right) \tag{6.49}$$

where $mean_e$ and var_e designate, respectively, the empirical mean and variance of d, over the set $P_2(\mathbf{E})$ of unordered distinct element pairs of $\mathbf{E}$. To be more explicit, we have

$$mean_e(D) = \frac{1}{\binom{m}{2}} \sum \{d(y, z)|\{y, z\} \in P_2(\mathbf{E})\} \tag{6.50}$$

$$var_e(D) = mean_e(D^2) - \left(mean_e(D)\right)^2 \tag{6.51}$$

where $m = card(\mathbf{E})$ and where $\binom{m}{2}$ is a binomial coefficient.

In these conditions, the table of the similarity indices (7.2.6) of Chap. 7 of [1] has to be taken up again. Let us recall here its expression:

$$\mathcal{P}^g = \{P^g(y, z) = \Phi(Q^g(y, z))|\{y, z\} \in P_2(\mathbf{E})\} \tag{6.52}$$

where Φ is the cumulative normal distribution function.

Therefore, as just mentioned above, by referring to Eqs. (10.4.7) and (10.4.8) of Sect. 10.4 of Chap. 10 of [1], for a given value ϵ of the LLA criterion ($0 \le \epsilon \le 1$), we can write

$$Crit_{LLA}(C_l^k, x) = -log_2\left(-log_2(LL_{max}^\epsilon(C_l^k, \{x\}))\right) \tag{6.53}$$

where

$$LL_{max}^\epsilon(C_l^k, \{x\}) = \left(p(C_l^k, \{x\})\right)^{\pi(l,k)} \tag{6.54}$$

where $\pi(l, k) = (card(C_l^k))^\epsilon$.

$$p(C_l^k, \{x\}) = max\{P^g(y, z)|y \in C_l^k\} \tag{6.55}$$

As above, let us insist on the fact that a partition into K clusters obtained by any of the *APMSR* methods permits the *K-means* algorithm to be initialized efficiently. This will be emphasized once more in Sect. 6.5, reserved to the developments [7, 8].

6.3 Criteria and Algorithmic of the Attraction Pole Methods by Successive Aggregations (*APMSA*)

6.3.1 Introduction

We shall express the algorithmic approach underlying this family of methods in the framework of clustering the set of column attributes or that of row objects of an incidence data table (see (2.1) and (2.2) of Chap. 2). In this context, to begin, we define two associated distance indices: the former being on the column attributes and the latter on the row objects. These indices correspond mutually. They can be used in place of the index designated by D in the development above of the *APMSR* methods.

As for *APMSR* methods (see Sect. 6.2), the data is defined by a set $\mathbf{E}$ provided with a distance or dissimilarity index d. An equivalent process can be considered in the case where instead of D, a similarity index S is given on $\mathbf{E}$. We have just specified above (see (6.11) and (6.12)) the distance indices concerned by clustering the set of column attributes and that of row objects.

The algorithm we shall define corresponds to a recurrent process. It turns out a single partition, class after class. Let us consider the determination of the first class. The latter will be driven by an attraction pole P established on the basis of the following distance distribution:

$$\{d(P, x) | x \in \mathbf{E} - \{P\}\} \tag{6.56}$$

The criterion employed for extracting P is the maximization of the variance or that of the absolute moment of order 2 of the (6.1) distribution.

In these conditions, according to (6.2), the increasing sequence of the distances to P can be designated as

$$d(P, x_{(1)}) \leq d(P, x_{(2)}) \leq \cdots \leq d(P, x_{(j)}) \leq \cdots \leq d(P, x_{(m-1)}) \tag{6.57}$$

We associate a positive real axis of origin P with these numerical values. In addition to P, $m - 1$ points whose abscissas are $x_{(1)}, x_{(2)} \ldots x_{(j)} \ldots$ and $x_{(m-1)}$ are considered.

One of the algorithmic versions of the formation process of a given class can be intuitively expressed as follows. We start by aggregating to the attraction pole P a sequence of $\mathbf{E}$ elements ordered by distance to P. The size order of this sequence has to be fixed. Thereby, necessarily, the set of the $\mathbf{E}$ elements joined to P has the following form:

$$\{x_{(j)} | 1 \leq j \leq min\} \tag{6.58}$$

where *min* is a bound we fix.

Therefore, the matter consists of stopping the aggregation till $x_{(k)}$ $(k > min)$, if the point whose abscissa is $d(P, x_k)$ may appear as a center of a *small* interval where the point density—in the linear cloud defined by (6.57)—is appreciably weaker on the right than on the left.

The technique used refers to the fitting of (6.57) by a Gaussian distribution. More precisely, to the sequence of distances to P (6.57) corresponds the sequence

$$\{\Phi_l | 1 \leq l \leq m - 1\}$$

where

$$\Phi_l = \Phi\Big(\frac{d(P, x_{(l)}) - \mu_l}{\sigma_l}\Big) \tag{6.59}$$

In this equation μ_l and σ_l designate, respectively, the mean and standard deviation of the distribution defined by the sequence (6.57). Φ indicates the normal cumulative distribution function.

In these conditions, the cluster formation is stopped to

$$\{P = x_{(0)}, x_{(1)}, x_{(2)}, \ldots, x_{(k)}\}$$

if an appreciable difference occurs between Φ_k and Φ_{k+1}, Φ_{k+1} being greater than Φ_k.

This method, very simple in its principle, has given very interesting results in certain situations. Nevertheless, some difficulties may appear in the algorithmic evaluation of the gap between Φ_k and Φ_{k+1}. For reasons of space and focus of the subject, these developments have not been reported here.

For this reason and also for generalization purpose of this type of approach, the assignment of an element $\mathbf{E}$ to a cluster C in its process of construction will be carried out by using a probabilistic threshold, that we will denote by $Prob^C$.

More precisely, if x_h is the latest $\mathbf{E}$ element to have integrated the cluster C already built around the pole P, $\Phi(d(x_{h+1}, P))$ is compared to $Prob^C$. Then, x_{h+1} is aggregated to C if

$$\Phi(d(x_{(h+1)}, P)) \leq Prob^C \tag{6.60}$$

where the parameters of the normal distribution (*mean* and *variance*) are calculated from the distribution defined by (6.45).

If this last inequality is not satisfied, the formation of the cluster concerned is stopped.

In this technique the association of an element to a cluster is based on the proximity of this element to the attraction pole P concerned by the cluster in the process of formation absorbed by P. In fact, the matter corresponds to an element of the method family *APMSA*.

The constitution of clusters can be used only for the determination of the poles. In these conditions, the clusters can be reconstituted using an *APMSR* family algorithm. That's what was done in Sect. 6.6.2 (see Fig. 6.9)

6.3.2 Criteria for APMSA

As seen in Sect. 6.2.3, different assignment criteria of an element to a class can be considered in the framework of the *APMSR* methods. This will be equally the case for the *APMSA* methods. Let us denote these criteria as follows:

$$P_0(C_l, x) \, , \, P_{SL}(C_l, x) \, , \, P_{CL}(C_l, x) \, , \, P_{AL}(C_l, x)$$
$$P_{PAL}(C_l, x) \, , \, P_W(C_l, x) \, , \, P_{LLA}(C_l, x) \tag{6.61}$$

where C_l is the only cluster already constituted around the attraction pole concerned, after $l - K$ assignments in all. l is the cardinality of the already clustered subset of **E**.

For a given criterion (there are seven in all), once the class has been completed around a given attraction pole (this because the proximity of every element not yet clustered to the class constituted, is lower than the threshold chosen), the entire process is reiterated on the set not yet clustered with the same probabilistic threshold.

Now, let us clarify how a given class is built step by step and how, in particular, the probabilistic proximity between an element—not yet classified—and the class concerned is evaluated.

At a given step of the algorithm progression, just after have completed a given class, let R designate the **E** subset not yet clustered. Initially, $R = \mathbf{E}$. R will provide us a class after the determination of an attraction pole within it and the choice of one of the criteria (6.61). Recall that the latter results from a probabilistic normalization of the criteria listed in (6.40).

The attraction pole in R, that we designate it here by Q, is determined from the variance (or the absolute moment of order 2) of the distance distribution in R (see 6.57). In order to be able to pass from one of the criteria (6.40) to its probabilistic version, we begin by aggregating to Q a minimal set composed of a few elements, those closest to Q in the sense of the distance d. We can, for example, take two or three elements for this minimal set. The pole surrounded by these few elements defines the initial state of the class to be built.

After that, one of the criteria (6.40) is chosen. Let us designate it by $Crit_A(C_l, \{x\})$. The latter compares the already class C_l formed around Q and the element x of R. Now, let us specify how the progression is made in the class formation. The matter is

"Is it necessary to integrate one additional element into the class concerned or to stop its formation?"

To answer, first, the internal distribution of the distances of an element of C_l to the other elements of C_l, namely,

$$\{Crit_A(C_l, \{x\}) | x \in C_l\} \tag{6.62}$$

is modeled. For this, we use the approximation defined by the probability normal law with the following mean and variance parameters:

$$mean_l = \frac{1}{l} \sum_{x \in C_l} Crit_A(C_l - \{x\}, \{x\})$$

$$var_l = \frac{1}{l} \sum_{x \in C_l} [Crit_A(C_l - \{x\}, \{x\}) - mean_l]^2 \qquad (6.63)$$

Consider now an element y of $R = E - C_l$, its probabilistic proximity to C_l is defined by

$$P_A(C_l, \{y\}) = \Phi(z) \qquad (6.64)$$

where

$$z = \frac{Crit_A(C_l, \{y\} - mean_l)}{\sqrt{var_l}} \qquad (6.65)$$

In these conditions, for an element y of R realizing the maximal value of $P_A(C_l, \{y\})$ (in most cases there is a unique such an element y), $P_A(C_l, \{y\})$ is compared to a probabilistic threshold $Prob^C$. And then, if $P_A(C_l, \{y\})$ is greater or equal to $Prob^C$, the element y integrates the cluster being set up. Otherwise, the class formation is stopped.

The adjustment of the value of the probabilistic threshold $Prob^C$ is a delicate problem. For this, it may be useful to compare the effect of distinct values, differentiated enough, on the clusterings obtained, respectively. Also, the use of a fitting criterion between a classification obtained on E and the distance d endowing E, may contribute efficiently to the comparison analysis.

For the latter, two criteria studied in ascendant hierarchical clustering can be considered: the adequation criterion developed in Sects. 9.3.5 and 9.3.6 of Chap. 9 of [1] and the Ward inertia criterion expressed in Sect. 10.3.3 of Chap. 10 of [1].

6.4 On the Number of Poles to Extract and on the Quality of the Clusterings Obtained

As mentioned in the last paragraph above, we will be inspired by the criteria studied in the framework of Ascendant Hierarchical Clustering. It is, on the one hand, the fitting criterion between a *Similarity* on the set E to be clustered and a partition of E. On the other hand, for an euclidean representation of E, it is the explained inertia criterion. Each of both criteria evaluates in a certain way, how much the partition of E approximates the *Similarity* or the *Distance* defined on E.

6.4.1 The Fitting Criterion

We refer here to a finite set $\mathbf{E}$ endowed with a similarity index S. The latter can be given by a similarity table as follows:

$$S = \{S(p)|p = \{x, y\} \in P_2(\mathbf{E})\} \tag{6.66}$$

where $P_2(\mathbf{E})$ is the set of all unordered distinct element pairs (2-subsets) of $\mathbf{E}$ and where $S(p)$ is the value of the similarity function S on the pair $\{x, y\}$.

If π designates the partition whose appropriateness has to be evaluated, the fitting criterion can be written in the following form:

$$CA(S, \pi) = \frac{1}{\sqrt{\left(\frac{r(\pi).s(\pi)}{f-1}\right)}} \cdot \sum_{p \in \mathbf{F}} \rho(p).c(p) \tag{6.67}$$

This form takes up again (4.3.70) of Sect. 9.3.52 of Chap. 9 of [1]. Now, let us recall explicitly its different components.

$\mathbf{F} = P_2(\mathbf{E})$ is the set of unordered pairs (or parts with two elements) of $\mathbf{E}$. f designates the cardinality of F. If m denotes the cardinal of $\mathbf{E}$, $f = m(m-1)/2$. $r(\pi) = card(R(\pi))$ and $s(\pi) = card(S(\pi))$ where $R(\pi)$ (resp., $S(\pi)$) is the set of joined pairs (resp., separated pairs) by the partition π ($r(\pi) + s(\pi) = f$). ρ denotes the indicator function of $R(\pi)$.

With these notations we have

$$\sum_{p \in \mathbf{F}} \rho(p) = r(\pi) \tag{6.68}$$

On the other hand,

$$(\forall p \in \mathbf{F}), c(p) = \frac{S(p) - mean_e(S)}{\sqrt{var_e(S)}} \tag{6.69}$$

defines the standardized similarity measure. In this, the centering and reducing $S(p)$ is performed empirically with respect to the observed distribution of the similarity function S, on the set F. More explicitly, we have

$$mean_e(S) = \frac{1}{f} \sum_{p \in \mathbf{F}} S(p)$$

and

$$var_e(S) = \frac{1}{f} \sum_{p \in \mathbf{F}} (S(p) - mean_e(S)) \tag{6.70}$$

In our development above we have placed the matter immediately in the case where the data is defined by a set $\mathbf{E}$ provided with a distance or dissimilarity index d (see, for example, Sect. 6.2.1). In these conditions, the passage to the (6.54) formulation of the criterion can be get by putting

$$(\forall p \in \mathbf{F}), \, S(x, y) = d_{max} - d(x, y)$$

where

$$d_{max} = max\{d(x, y)|\{x, y\} \in \mathbf{F}\} \tag{6.71}$$

Now, in the case where $\mathbf{E}$ is defined by the set $\mathcal{A}$ of the column attributes and also, where the square of the distance concerned is $D_{\alpha}^2(k, l)$ (see (6.11)), the similarity matrix for the $S(k, l)$ can be given as it is specified in (6.9).

Similarly, in the case where $\mathbf{E}$ corresponds to the set of row objects $\mathcal{O}$ and where the square of the distance concerned is $D_{\omega}^2(h, i)$ (see (6.13)) the similarity matrix can be given by

$$P(h, i) = \frac{1}{2}\{-D_{\omega}^2(h, i) + P(i, i) + P(h, h)\} \tag{6.72}$$

where for a row object i' we have:

$$P(i, i') = \frac{1}{\sqrt{n}} \sum_{1 \leq j \leq p} \frac{(x_{i'}^j - \eta_j)^2}{\eta_j} \tag{6.73}$$

6.4.2 Inertia or Ward Criterion

Designate by

$$\pi(\mathbf{E}) = \{E_k | 1 \leq k \leq K\} \tag{6.74}$$

the partition of $\mathbf{E}$ to be evaluated. We set $m_k = card\,(E_k)$ so that

$$m = \sum_{1 \leq k \leq K} m_k$$

$\mathbf{E}$ is supposed to be endowed with a distance index D. The latter can derive from an euclidean representation (see 6.2.1).

To simplify—but generalization is straightforward—we assume the elements of $\mathbf{E}$ equally weighted.

Now, let E_k expressed as

$$E_k = \{e_{ki} | 1 \leq i \leq m_k\} \tag{6.75}$$

for $k = 1, 2, \ldots, K$.

The inertia formulation we carry is that defined from (10.3.5) of Sect. 10.3.3. of Chap. 10 of [1]. It is associated with a cloud of points:

$$\mathcal{N}(\mathbb{I}) = \{(O_i, \mu_i) | i \in \mathbb{I}\}$$

In these conditions, the total inertia moment can be written as follows:

$$\frac{1}{2\mu} \sum_{(i,i') \in \mathbb{I} \times \mathbb{I}} d^2(O_i, O_{i'})$$

And then, the retained inertia by the partition $\pi(E)$ can be written as

$$W(\pi(E)) = \sum_{1 \leq k \leq K} \frac{1}{2m_k} \sum_{1 \leq i, i' \leq m_k} d^2(e_{ki}, e_{ki'}) \tag{6.76}$$

Therefore, the ratio of the explained inertia by the partition $\pi(E)$ becomes

$$\frac{W(\pi(E))}{\frac{1}{2m} \sum_{1 \leq i, i' \leq m} d^2(e_i, e_{i'})} \tag{6.77}$$

This expression can be reduced to that of $i(\pi(E))$ specified as follows:

$$i(\pi(E)) = m. \sum_{1 \leq k \leq K} \frac{1}{m_k} \cdot \frac{Sum(d^2(E_k))}{Sum(d^2(E))} \tag{6.78}$$

where Sum is extended over all the ordered pairs of the argument concerned. This formula is proposed in [9].

Now, we suppose that the set $\mathbf{E}$ is either the set $\mathcal{A}$ of the column attributes or the set $\mathcal{O}$ of the row objects. On the other hand suppose given a similarity matrix Q on $\mathbf{E}$ that we designate by $Q(E)$. Let $Sum(Q(E))$ and $Trace(Q(E))$, respectively, the sum of all the values and the sum of diagonal values in $Q(E)$.

A calculation developed in [9] shows that for $Q = S$ and $E = \mathcal{A}$ or for $Q = P$ and $E = \mathcal{O}$ (see (6.9) and (6.10)) the part of the inertia explained by the partition $\pi(E)$ can be expressed ac follows:

$$i(\pi(E)) = \frac{\sum_{1 \leq k \leq K} \frac{1}{card(E_k)} \cdot Sum(Q(E_k)) - \frac{1}{m} \cdot Sum(Q(E))}{Trace(Q(E)) - \frac{1}{m} \cdot Sum(Q(E))} \tag{6.79}$$

6.4.3 Interesting Partitions from APMSR and from APMSA

Each element of the *APMSR* algorithm family is associated with a criterion of the form $Crit(C_l^k, x)$ (see (6.40)). C_l^k is the cluster already constituted around the kth pole after $l - K$ assignments exactly. In this case, the number of classified elements is exactly l. x is an element of the subset R_l of $\mathbf{E}$ defined by the subset not yet classified after $l - K$ assignments (see (6.16)). Therefore, $card(R_l) = m - l$.

An algorithm of *APMSR* family produces a sequence of descendant but non-hierarchical classifications. Note $2, 3, \ldots, h, \ldots, H$, the respective numbers of classes of the partition sequence get. The latter according to (6.61) can be put in the form:

$$\Xi_H = \{\pi_h(E) | 2 \leq h \leq H\} \tag{6.80}$$

where $\pi_h(E)$ is a partition of $\mathbf{E}$ around h attraction poles, each of them attracting one of the classes.

To begin, in agreement with the expert, a value of H is fixed. On the other hand, one of the seven criteria presented in (6.40) is adopted. Recall here that a given criterion among these is defined by an association coefficient between an element not yet classified and a cluster in its construction process. Finally, a determination technique of a set of attraction poles is retained (see Sect. 6.2.2). Thereby, we end by obtaining a partition sequence Ξ_H (see (6.80)).

In these conditions, let us be inspired by the method of determining the most coherent partitions or nodes that can be collected at the various levels of a hierarchical clustering (see Sect. 9.3.6 of Chap. 9 of [1]).

The argument of the method is an association or proximity coefficient between a partition of $\mathbf{E}$ and a similarity S, or a distance index d on the set $\mathbf{E}$.

We shall consider two different types of such an association coefficient, called—in the context concerned—evaluation criterion of a partition. The first one is the coefficient $CA(S, \pi)$ (see (6.67)) and the second one is $W(\pi(E))$ (see (6.76)).

The strategy will be the same for each of both criteria. Let us consider the first of them $CA(S, \pi)$.

We will associate with the partition sequence Ξ_H (see (6.80)), the ordered sequence of the values of the criterion concerned, that is

$$CA(S, \Xi_H) = \big(CA(S, \pi_h(E)) | 2 \leq h \leq H\big) \tag{6.81}$$

where $CA(S, \pi_h(E))$ is the value of the adequation coefficient (6.67) between the similarity S and the partition $\pi_h(E)$ (into h classes around h poles).

In the partition sequence Ξ_H, we seek to determine those of the partitions for which $CA(S, \pi_h(E))$ is large enough. It is of importance that the partitions (classifications, clusterings) retained are quite different, particularly in terms of the respective numbers of clusters.

Simultaneously, to the sequence of values $CA(S, \Xi_H)$ of the adequation criterion, we can consider the sequence of the variation rate of the latter, namely,

$$\Delta\big(CA(S, \Xi_H)\big) = \big(CA(S, \pi_h(E)) - CA(S, \pi_{h-1}(E))|2 \leq h \leq H\big) \qquad (6.82)$$

where, clearly, $\pi_1(E)$ is the rough partition reduced to a single class.

In the case of the second criterion based on the explained inertia, we consider the ordered sequence of the criterion values $W(\pi(E))$ (see (6.76)), that is

$$W(d, \Xi_H) = \big(W(d, \pi_h(E))|2 \leq h \leq H\big) \qquad (6.83)$$

As for the preceding criterion, we consider the sequence of the variation rate of W between two consecutive partitions, that is

$$\Delta\big(W(S, \Xi_H)\big) = \big(W(d, \pi_h(E)) - W(d, \pi_{h-1}(E))|2 \leq h \leq H\big) \qquad (6.84)$$

being heard that $\pi_1(E)$ is the rough partition including a single class.

6.5   Developments

There are two principal families in the clustering algorithms. The first one is developed in the context of ascendant hierarchical clustering [10, 11] and the second one, in the K-means methodology [2, 3]. Historically, these two families have been the most worked and the most applied. These two approaches and their respective scopes are clearly described in [1].

As the K-means, the attraction pole methods *APMSR* and *APMSA* are essentially non hierarchical clustering methods. Nevertheless, especially *APMSR*, they can carry out a sequence of partitions. They have an original and very specific nature with respect to the both main approaches mentioned.

The attraction pole clustering methods emerged at the end of the seventies [6, 9]. Their specificities and independent circumstances have made that they could not be distributed sufficiently.

Even so, the two types of algorithms defined by the K-means and the Ascendant Hierarchical Clustering can be added to *APMSR* and *APMSA* algorithms, respectively.

Now, let us make clear the K-means extension of a given method of the *APMSR* family. We get a partition

$$\pi(\mathbf{E}) = \{E_k|1 \leq k \leq K\} \qquad (6.85)$$

around K attraction poles by using one of the criteria (6.40), to each class we will associate an attraction *center* G_k, $1 \leq k \leq K$. If an euclidean representation (cloud of points in a geometrical space) can be considered for $\mathbf{E}$, G_k may be the center of gravity of the E_k representation, $1 \leq k \leq K$. Notice that this representation holds when $\mathbf{E}$ is associated with the set of column attributes (resp., row objects) of an incidence data table. Otherwise, G_k can be defined as the element of E_k minimizing the square sum of the distances to the other elements of E_k. More precisely we have

$$G_k = Arg\left\{ min_{g \in E_k}\left\{ \sum \{d^2(g, e_{ki})|1 \leq i \leq m_k\}\right\}\right\} \qquad (6.86)$$

where the e_{ki} are the elements of E_k (see (6.75)), $1 \leq k \leq K$ and where d is the distance considered (D_α or D_ω).

Experiment on real or simulated data shows that the K-means algorithm convergence is reached after two, and sometimes three iterations [9].

Moreover, theoretically, we analyze in [7] the behavior of the K-means algorithm in the case of a specific mathematical model corresponding to a perfect classifiability (see Sect. 8.7 of Chap. 8 of [1]). In the latter model, the set is composed of two consecutive intervals of an horizontal axis, I_1 and I_2, uniformly weighted with a density equal to 1, separated by an empty interval I. The respective lengths of I_1, I_2 and I are the parameters of the analysis of the K-means convergence algorithm. The optimal solution of the convergence is defined by an ordered pair (G_1, G_2) of attraction centers, where G_1 is the middle of I_1 and G_2 is the middle of I_2. We show that if the initial system of attraction centers is formed of the two first attraction poles p_1 and p_2, then, under fairly general conditions, the convergence of the K-means (here 2-means) algorithm to the optimal solution is acquired. An interesting and exhaustive analysis of our strategy is given in [8].

Now, relative to the *APMSA* algorithm a derivation in terms of hierarchical clustering occurs naturally. To see that, refer to the formalization of an ultrametric space in terms of a hierarchical family of circular spheres (see Sect. 1.4.2 of Chap. 1 of [1]). In these conditions, in order to establish the subclasses of a given class determined—at the level of the entire set—from the threshold $Prob^C$ (see (6.59)), we proceed—at the level of the class concerned—in the same way as it was at the level of the whole set. For simplicity and coherence reasons, we propose to use the same threshold $Prob^C$. However, now, the parameters of the distribution function Φ, as indicated in (6.59), are calculated in the framework of the class concerned. Clearly, for the construction of a total clustering hierarchy, the process will have a recursive character. It is important to notice that here, it will be a *descending* and not *ascending* construction of a hierarchy of classifications.

Now, it is essential to compare the respective behaviors of the different methods described above when there are applied to real or simulated data. We will attempt to do so in the following section.

6.6 Applying Clustering Attraction Pole Methods in Real Data

6.6.1 Introduction

As already expressed in the Preamble of Sect. 2.4.6 of Chap. 2, these are metric concepts highlighting mutual differences or oppositions that must be used in clustering by attraction pole methods. The notion of similarity or coefficient of association refers to a relative neutrality or independence between attributes and is therefore more oriented towards geometric representations and seriations (see Sect. 2.3.3 of Chap. 2 and Chaps. 4 and 5).

In these conditions, the distance squared between unit elements is D_α^2 if the set concerned is the set of attributes and D_ω^2 if it is the set of objects (see (6.11) and (6.12)).

These indices were established in the case where the descriptive attributes are Boolean. Their adaptation in the case of numerical attributes can be deduced (Chaps. 5 and 7 of [1] give the tools for such an extension).

Three classic and very popular datasets will be experimented : the Ruspini data (see Sect. 6.6.2), the Fisher data (see Sect. 6.6.3) and the Merovingian belt buckles-plates data (see Sect. 6.6.4).

Two related objectives result from these treatments. First of all, we must realize the respective behaviors of the different algorithms and criteria that we have introduced. Second and related, we need to submit our results to the data expert.

As we just said, Ruspini data [12] is very popular. They are generally used to test non-hierarchical clustering methods on a set of objects described by numerical variables. These data are defined by a cloud of 75 points in the geometric plane. Therefore, this illustrates the description of a set of 75 objects by two numerical variables. The distribution of points in the plane makes it possible to visually recognize four clusters (see Fig. 6.1). These are differently dense or fuzzy. Also, they are of different sizes and also, mutually, differently distant from each other. Precisely, these data were carried out to test a clustering approach based on the notion of fuzzy Zadeh membership of an object to a set [13, 14].

These data as well as the following two that we will mention below will be used to test the behavior of our classification algorithms of pole of attraction methods. The family of methods most concerned is that of *APMSR* by successive reassignments (see 6.2). Nevertheless, an example will be given of the application of a successive aggregation family *APMSA*.

Ruspini's data was particularly used to validate Dynamic Clustering method (see Chap. 2 of [1]). Therefore, we will indicate to conclude the comparison which was made between our new algorithms and that of dynamic clustering.

Fisher's Iris data is also very famous and very popular. The work done on them has a premonitory character in the analysis of multidimensional data [15]. The objective is to observe and then characterize the morphological differences between three species of iris flowers: iris setosa, iris virginica and iris versicolor. Four numerical

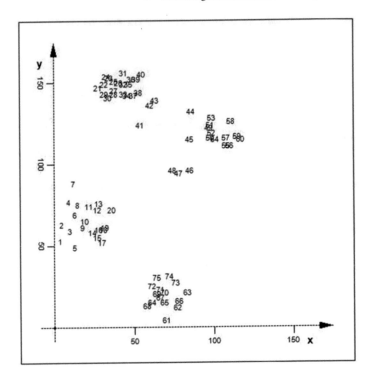

Fig. 6.1 The Ruspini Data

attributes are considered for description: sepal length, sepal width, petal length, and petal width.

Three independent samples, each consisting of 50 independent elements of the same species were taken. Thus, the data table as we usually consider it Objects x Variables, includes 150 rows (50 per species) and 4 columns.

The problem posed is in fact of supervised classification. It is a question of determining if the three species are differentiated and in what way they do it. Moreover, the method proposed by Fisher [15] is that of discriminant linear analysis [16].

However, these data have been used very frequently to test the ability of unsupervised classification methods—usually non-hierarchical—to separate the three species. It is in this context that we will evaluate the behavior of our algorithms based on the prior determination of attraction poles.

In the case of Merovingian belt buckles-plates, the description is defined by an incidence data table whose attributes are Boolean. Previously, the results of the treatment of two families of methods have been reported. In the first one (see Sect. 4.4.3 of Chap. 4) planar geometric representations are produced. These are provided from multidimensional data analysis methods. In the second family (see Sect. 5.4.3 of 5) combinatorial seriation methods are applied.

By presenting the results of the first family of methods, we have already indicated that both synthesis structures that are order and classification (seriation and clustering) necessarily influence the results of a data analysis regardless of its orientation.

In this chapter, the focus is clearly classification (clustering). We will note the nature and the quality of the results of the pole attraction methods. Comparison with those of hierarchical clustering is observed. Moreover, one will note with interest the difference in nature of these results with those obtained by seriation in Chap. 5.

6.6.2 Ruspini Data

We have already presented Ruspini's data in the introduction above (see Sect. 6.6.1). Figure 6.1 gives the representation of the concerned 75 points of the plane. We will show a set of classificatory organization results of these points, respectively, obtained by the algorithms that we have defined (see 6.2.3). Mainly, we will use methods of the *APMSR* family (see Sect. 6.2). These are distance criteria that will be used (6.40). They will be noted Alg-0, Alg-SL, Alg-AL, Alg-PAL, Alg-W and Alg-LLA (see Sect. 6.2.4.3).

In each of the following graphical representations, each pole drives a class (a cluster). From each of the poles emerge lines connecting the pole to the different elements of the class that it has trained. Four clusters—the same—have been obtained by each of the algorithms Alg-0, Alg-SL, Alg-AL, Alg-PAL and Alg-LLA (see Fig. 6.2). However, the order of aggregation at a given pole may not be the same regardless of the algorithm used. For this classification, the percentage of the explained inertia is 93.18%.

We can notice that the partition into four classes provided by Alg-W is slightly different from the previous one (see Fig. 6.3). In this, the percentage of the explained inertia is 88.68%.

In the enumeration of the four clusters represented in Fig. 6.2, the pole has been shown in bold and the gravity center of the class concerned in italics (Fig. 6.10).

Otherwise, partitions into five clusters were built from the same poles of attraction (see Figs. 6.4, 6.5 and 6.6). Interesting variants can be observed for each of them.

The algorithms concerned in Fig. 6.4 are Alg-SL, Alg-AL and Alg-W. The percentage of the explained inertia is 94.74%.

The algorithms concerned in Fig. 6.5 are Alg-0 and Alg-PAL. The percentage of the explained inertia is 94.38%.

The algorithm concerned in Fig. 6.6 is Alg-LLA. The percentage of the explained inertia is 94.41%.

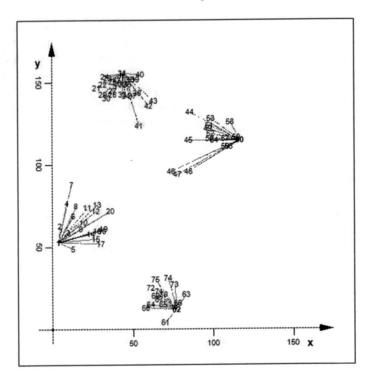

Fig. 6.2 The four clusters

Going up to seven classes per partition, the results are very slightly distinct. Nevertheless, they keep all their relevance because of the efficiency of the pole extraction method to build the clusters (see Figs. 6.7 and 6.8).

The result of Fig. 6.9 corresponds to a version of the *APMSA* approach comprising two phases where the second uses the *APMSR* approach (see Sect. 6.3). Recall that this version includes two stages. The objective of the first stage is to determine a set of poles in fixed number. The first pole determines a class relative to the total set. The second pole is obtained by the same way, in relation to the remaining set by having emptied the first class and so on ...

Once the set of poles has been obtained, the different elements of the total set—apart from the poles—are reassigned to the different poles.

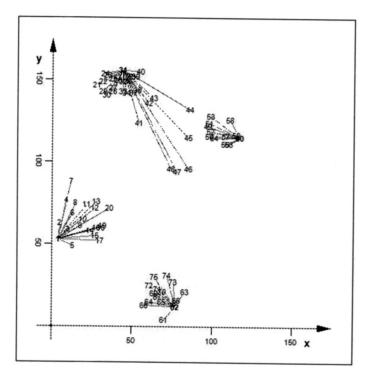

Fig. 6.3 The four clusters obtained by Alg-W

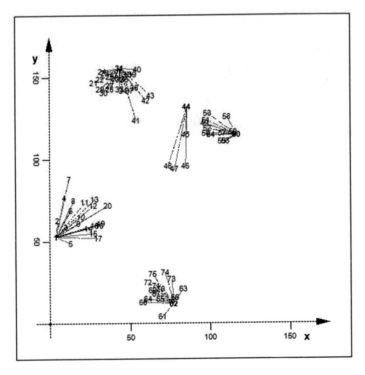

Fig. 6.4 Partition into 5 clusters, Alg-SL, Alg-AL and Alg-W

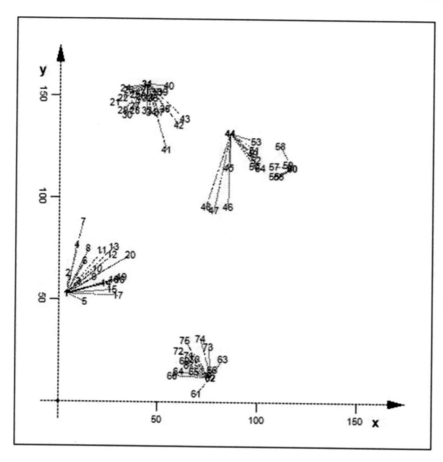

Fig. 6.5 Partition into 5 clusters, Alg-0 and Alg-PAL

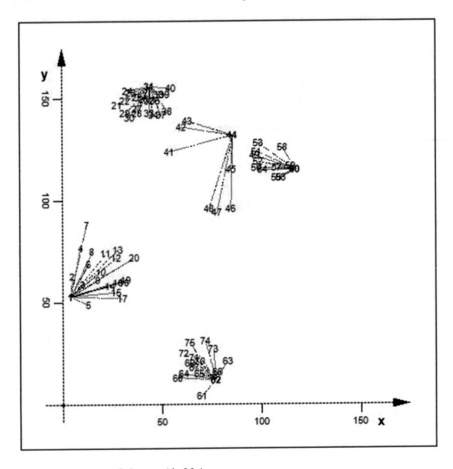

Fig. 6.6 Partition into 5 clusters, Alg-LLA

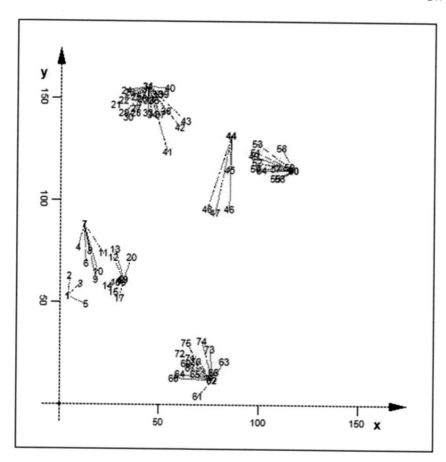

Fig. 6.7   Partition into 7 clusters, Alg-AL

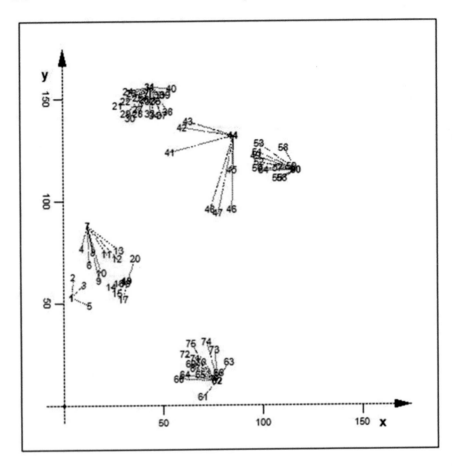

Fig. 6.8 Partition into 7 clusters, Alg-LLA

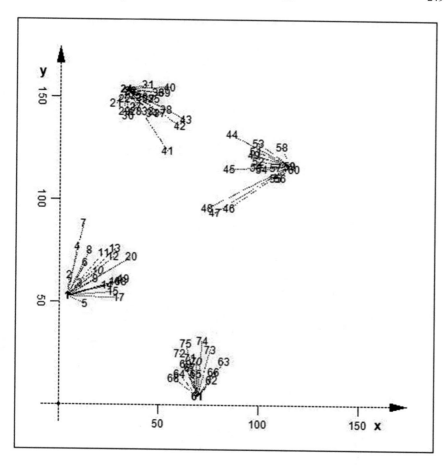

Fig. 6.9 Partition into 4 clusters, *APMSA* → *APMSR*, Alg-0 followed by Alg-AL

Partition into 4 clusters			
percentage of inertia explained : 93.18%			
criterion of the adequation quality : 38.86			
cluster 1	cluster 2	cluster 3	cluster 4
1	60	31	62
2	44	21	61
3	45	22	63
4	46	23	64
5	47	24	65
6	48	25	66
7	49	26	67
8	50	27	68
9	51	28	69
10	52	29	70
11	53	30	71
12	54	32	72
13	55	33	73
14	56	34	74
15	57	35	75
16	58	36	
17	59	37	
18		38	
19		39	
20		40	
		41	
		42	
		43	

Partition into 4 clusters			
percentage of inertia explained : 93.18%			
criterion of the adequation quality : 38.86			
cluster 1	cluster 2	cluster 3	cluster 4
1	60	31	62
2	59	32	66
3	56	26	63
5	55	35	73
6	57	36	70
9	58	33	65
10	54	34	74
8	52	37	67
4	50	39	71
11	51	38	69
14	49	25	75
12	53	27	72
16	45	28	64
15	44	23	61
13	46	30	68
18	47	24	
19	48	22	
17		29	
20		21	
7		40	
		42	
		41	
		43	

Partition into 4 clusters			
percentage of inertia explained : 93.18%			
criterion of the adequation quallity : 38.86			
cluster 1	cluster 2	cluster 3	cluster 4
1	60	31	62
2	59	32	66
3	56	35	63
6	55	36	73
8	57	26	74
4	58	39	70
10	54	25	71
9	50	33	67
11	52	34	69
12	49	27	65
13	51	28	72
14	53	23	75
15	45	24	64
16	44	22	68
18	46	30	61
19	47	29	
17	48	37	
5		38	
20		21	
7		40	
		42	
		43	
		41	

Partition into 4 clusters			
percentage of inertia explained : 93.18%			
criterion of the adequation quallity : 38.86			
cluster 1	cluster 2	cluster 3	cluster 4
1	60	31	62
2	59	32	66
3	56	35	63
5	55	36	73
6	57	39	74
8	58	26	70
4	54	25	71
10	50	23	69
9	52	24	67
14	49	22	65
15	51	40	64
16	53	29	68
18	44	30	72
19	45	33	75
17	46	34	61
11	47	27	
12	48	28	
13		37	
7		38	
20		21	
		42	
		43	
		41	

Fig. 6.10 Enumeration of the four clusters

6.6.3 Fisher Data

Fisher's Iris data was introduced in Sect. 6.6.1. The reader is asked to reconsider this introduction.

In each of the three tables results (see Fig. 6.11) The class of Iris setosa is perfectly separated. By using the Alg-PAL algorithm, this separation result is almost equally observed for Iris virginica on one side of the table as for Iris versicolor on the other

Partition into 3 clusters		
percentage of inertia explained : 76.66% criterion of the adequation quality : 73.94		
cluster 1	cluster 2	cluster 3
119 virginica	23 setosa	61 versicolor
123 virginica	38 setosa	94 versicolor
131 virginica	5 setosa	58 versicolor
108 virginica	41 setosa	99 versicolor
106 virginica	1 setosa	82 versicolor
136 virginica	18 setosa	81 versicolor
103 virginica	28 setosa	54 versicolor
130 virginica	8 setosa	70 versicolor
126 virginica	40 setosa	90 versicolor
140 virginica	29 setosa	80 versicolor
113 virginica	27 setosa	91 versicolor
121 virginica	12 setosa	93 versicolor
142 virginica	50 setosa	95 versicolor
144 virginica	25 setosa	83 versicolor
141 virginica	7 setosa	68 versicolor
146 virginica	44 setosa	60 versicolor
105 virginica	24 setosa	100 versicolor
125 virginica	21 setosa	56 versicolor
148 virginica	32 setosa	97 versicolor
117 virginica	37 setosa	65 versicolor
78 versicolor	36 setosa	72 versicolor
111 virginica	22 setosa	89 versicolor
138 virginica	49 setosa	96 versicolor
116 virginica	11 setosa	74 versicolor
145 virginica	20 setosa	79 versicolor
129 virginica	47 setosa	67 versicolor
133 virginica	45 setosa	62 versicolor
104 virginica	30 setosa	64 versicolor
53 versicolor	3 setosa	98 versicolor
87 versicolor	48 setosa	85 versicolor
112 virginica	35 setosa	92 versicolor
128 virginica	10 setosa	84 versicolor
77 versicolor	31 setosa	75 versicolor
66 versicolor	4 setosa	134 virginica
76 versicolor	43 setosa	127 virginica
51 versicolor	2 setosa	55 versicolor
52 ver sicolo	26 setosa	139 virginica
57 versicolor	46 setosa	102 virginica
101 virginica	13 setosa	143 virginica
137 virginica	39 setosa	124 virginica
149 virginica	14 setosa	135 virginica
71 versicolor	9 setosa	73 versicolor
109 virginica	17 setosa	150 virginica
86 versicolor	6 setosa	122 virginica
115 virginica	19 setosa	59 versicolor
110 virginica	33 setosa	114 virginica
118 virginica	15 setosa	147 virginica
132 virginica	34 setosa	88 versicolor
	16 setosa	120 virginica
	42 setosa	69 versicolor
		63 versicolor
		107 virginica

Partition into 3 clusters		
percentage of inertia explained : 73.70% criterion of the adequation quality : 71.32		
cluster 1	cluster 2	cluster 3
119 virginica	23 setosa	61 versicolor
123 virginica	38 setosa	94 versicolor
106 virginica	5 setosa	58 versicolor
136 virginica	1 setosa	99 versicolor
131 virginica	18 setosa	82 versicolor
108 virginica	41 setosa	81 versicolor
130 virginica	28 setosa	70 versicolor
103 virginica	40 setosa	90 versicolor
113 virginico	8 setosa	54 versicolor
140 virginica	29 setosa	91 versicolor
142 virginica	27 setosa	93 versicolor
146 virginica	50 setosa	83 versicolor
141 virginica	12 setosa	95 versicolor
121 virginica	25 setosa	100 versicolor
144 virginica	21 setosa	56 versicolor
105 virginica	32 setosa	97 versicolor
148 virginica	24 setosa	68 versicolor
125 virginica	37 setosa	80 versicolor
145 virginica	36 setosa	96 versicolor
117 virginica	7 setosa	89 versicolor
138 virginica	44 setosa	65 versicolor
111 virginica	22 setosa	67 versicolor
116 virginica	49 setosa	85 versicolor
78 versicolor	11 setosa	62 versicolor
104 virginica	20 setosa	79 versicolor
129 virginica	47 setosa	64 versicolor
133 virginica	45 setosa	92 versicolor
112 virginica	3 setosa	98 versicolor
124 virginica	48 setosa	72 versicolor
127 virginica	30 setosa	74 versicolor
128 virginica	4 setosa	75 versicolor
139 virginica	31 setosa	59 versicolor
150 virginica	35 setosa	76 versicolor
84 versicolor	10 setosa	55 versicolor
71 versicolor	2 setosa	134 virginica
126 virginica	13 setosa	66 versicolor
102 virginica	26 setosa	87 versicolor
143 virginica	46 setosa	53 versicolor
122 virginica	43 setosa	77 versicolor
147 virginica	39 setosa	52 versicolor
101 virginica	14 setosa	51 versicolor
137 virginica	9 setosa	60 versicolor
149 virginica	6 setosa	57 versicolor
114 virginica	17 setosa	135 virginica
115 virginica	19 setosa	73 versicolor
109 virginica	15 setosa	86 versicolor
110 virginica	33 setosa	88 versicolor
118 virginica	34 setosa	69 versicolor
132 virginica	16 setosa	120 virginica
	42 setosa	63 versicolor
		107 virginica

Partition into 3 clusters		
percentage of inertia explained : 74.16% criterion of the adequation quality : 72.52		
cluster 1	cluster 2	cluster 3
119 virginica	23 setosa	61 versicolor
123 virginica	38 setosa	94 versicolor
106 virginica	5 setosa	58 versicolor
136 virginica	1 setosa	99 versicolor
131 virginica	18 setosa	82 versicolor
108 virginica	28 setosa	81 versicolor
130 virginica	41 setosa	70 versicolor
126 virginica	29 setosa	90 versicolor
103 virginica	40 setosa	91 versicolor
113 virginica	8 setosa	95 versicolor
140 virginica	50 setosa	100 versicolor
142 virginica	12 setosa	56 versicolor
146 virginica	25 setosa	97 versicolor
141 virginica	36 setosa	96 versicolor
121 virginica	27 setosa	89 versicolor
125 virginica	21 setosa	83 versicolor
145 virginica	32 setosa	93 versicolor
105 virginica	24 setosa	68 versicolor
148 virginica	7 setosa	67 versicolor
117 virginica	10 setosa	65 versicolor
138 virginica	31 setosa	54 versicolor
104 virginica	2 setosa	80 versicolor
78 versicolor	26 setosa	62 versicolor
87 versicolor	13 setosa	79 versicolor
66 versicolor	46 setosa	64 versicolor
76 versicolor	4 setosa	92 versicolor
53 versicolor	48 setosa	98 versicolor
59 versicolor	3 setosa	75 versicolor
51 versicolor	30 setosa	72 versicolor
55 versicolor	43 setosa	74 versicolor
134 virginica	37 setosa	139 virginica
77 versicolor	44 setosa	128 virginica
111 virginica	39 setosa	150 virginica
116 virginica	14 setosa	71 versicolor
52 versicolor	9 setosa	127 virginica
57 versicolor	22 setosa	124 virginica
86 versicolor	20 setosa	112 virginica
84 versicolor	47 setosa	129 virginica
102 virginica	45 setosa	133 virginica
143 virginica	49 setosa	147 virginica
122 virginica	11 setosa	73 versicolor
135 virginica	6 setosa	60 versicolor
114 virginica	17 setosa	88 versicolor
101 virginica	19 setosa	69 versicolor
137 virginica	15 setosa	120 virginica
149 virginica	34 setosa	63 versicolor
115 virginica	33 setosa	109 virginica
110 virginica	16 setosa	107 virginica
118 virginica	42 setosa	
132 virginica		

Fig. 6.11 Three clusters obtained by Alg-AL, Alg-PAL and Alg-LLA, respectively

side. By using Alg-AL and Alg-LLA, there is at the bottom of the columns, a mixture of Iris virginica and Iris versicolor. This result is natural given their morphological proximity. Moreover, the explained inertia for these new classifications are a little higher than in the previous case (use of Alg-PAL).

Considering partitions with a higher number of classes (5 or 6 classes) (see Figs. 6.12 and 6.13) and using, for example, the Alg-PAL algorithm, we see that the different types of Iris are quite isolated and that there is always an aggregate with a mixture of Iris virginica and Iris versicolor.

Partition into 5 clusters				
percentage of inertia explained : 80.66%				
criterion of the adequation quallity : 64.40				
cluster 1	cluster 2	cluster 3	cluster 4	cluster 5
119 virginica	23 setosa	61 versicolor	118 virginica	115 virginica
123 virginica	38 setosa	94 versicolor	132 virginica	122 virginica
106 virginica	5 setosa	58 versicolor	110 virginica	102 virginica
136 virginica	1 setosa	99 versicolor	145 virginica	143 virginica
131 virginica	18 setosa	82 versicolor	125 virginica	84 versicolor
108 virginica	41 setosa	81 versicolor	116 virginica	127 virginica
130 virginica	28 setosa	70 versicolor	111 virginica	124 virginica
103 virginica	40 setosa	90 versicolor	137 virginica	112 virginica
113 virginica	8 setosa	54 versicolor	149 virginica	134 virginica
140 virginica	29 setosa	91 versicolor	101 virginica	55 versicolor
142 virginica	27 setosa	93 versicolor	138 virginica	64 versicolor
146 virginica	50 setosa	83 versicolor	117 virginica	79 versicolor
141 virginica	12 setosa	95 versicolor	148 virginica	92 versicolor
121 virginica	25 setosa	100 versicolor	104 virginica	98 versicolor
144 virginica	21 setosa	56 versicolor	78 versicolor	75 versicolor
105 virginica	32 setosa	97 versicolor	128 virginica	74 versicolor
126 virginica	24 setosa	68 versicolor	139 virginica	72 versicolor
109 virginica	37 setosa	80 versicolor	150 virginica	62 versicolor
	36 setosa	96 versicolor	71 versicolor	59 versicolor
	7 setosa	89 versicolor	129 virginica	76 versicolor
	44 setosa	65 versicolor	133 virginica	66 versicolor
	22 setosa	67 versicolor	86 versicolor	87 versicolor
	49 setosa	85 versicolor		53 versicolor
	11 setosa	60 versicolor		77 versicolor
	20 setosa	63 versicolor		51 versicolor
	47 setosa	88 versicolor		52 ver sicolo
	45 setosa	69 versicolor		57 versicolor
	3 setosa	120 virginica		135 virginica
	48 setosa	107 virginica		73 versicolor
	30 setosa	42 setosa		147 virginica
	4 setosa			114 virginica
	31 setosa			
	35 setosa			
	10 setosa			
	2 setosa			
	13 setosa			
	26 setosa			
	46 setosa			
	43 setosa			
	39 setosa			
	14 setosa			
	9 setosa			
	6 setosa			
	17 setosa			
	19 setosa			
	15 setosa			
	33 setosa			
	34 setosa			
	16 setosa			

Fig. 6.12 Five clusters obtained by Alg-AL

Partition into 6 clusters					
percentage of inertia explained : 84.47%					
criterion of the adequation quallity : 51.01					
cluster 1	cluster 2	cluster 3	cluster 4	cluster 5	cluster 6
119 virginica	23 setosa	61 versicolor	118 virginica	115 virginica	16 setosa
123 virginica	38 setosa	94 versicolor	132 virginica	122 virginica	34 setosa
106 virginica	5 setosa	58 versicolor	110 virginica	102 virginica	33 setosa
136 virginica	1 setosa	99 versicolor	145 virginica	143 virginica	15 setosa
131 virginica	18 setosa	82 versicolor	125 virginica	84 versicolor	17 setosa
108 virginica	41 setosa	81 versicolor	116 virginica	127 virginica	6 setosa
130 virginica	28 setosa	70 versicolor	111 virginica	124 virginica	19 setosa
103 virginica	40 setosa	90 versicolor	137 virginica	112 virginica	20 setosa
113 virginica	8 setosa	54 versicolor	149 virginica	134 virginica	47 setosa
140 virginica	29 setosa	91 versicolor	101 virginica	55 versicolor	45 setosa
142 virginica	27 setosa	93 versicolor	138 virginica	64 versicolor	22 setosa
146 virginica	50 setosa	83 versicolor	117 virginica	79 versicolor	49 setosa
141 virginica	12 setosa	95 versicolor	148 virginica	92 versicolor	11 setosa
121 virginica	25 setosa	100 versicolor	104 virginica	98 versicolor	
144 virginica	21 setosa	56 versicolor	78 versicolor	75 versicolor	
105 virginica	32 setosa	97 versicolor	128 virginica	74 versicolor	
126 virginica	24 setosa	68 versicolor	139 virginica	72 versicolor	
109 virginica	37 setosa	80 versicolor	150 virginica	62 versicolor	
	36 setosa	96 versicolor	71 versicolor	59 versicolor	
	7 setosa	89 versicolor	129 virginica	76 versicolor	
	44 setosa	65 versicolor	133 virginica	66 versicolor	
	3 setosa	67 versicolor	86 versicolor	87 versicolor	
	48 setosa	85 versicolor		53 versicolor	
	30 setosa	60 versicolor		77 versicolor	
	4 setosa	63 versicolor		51 versicolor	
	31 setosa	88 versicolor		52 versicolo	
	35 setosa	69 versicolor		57 versicolor	
	10 setosa	120 virginica		135 virginica	
	2 setosa	107 virginica		73 versicolor	
	13 setosa	42 setosa		147 virginica	
	26 setosa			114 virginica	
	46 setosa				
	43 setosa				
	39 setosa				
	14 setosa				
	9 setosa				

Fig. 6.13 Six clusters obtained by Alg-PAL

6.6.4   Merovingian Belt Buckles-Plates Data

Let us take again the example of Merovingian buckles-plates considered in Sect. 4.4.3.1 of Chap. 4 and in Sect. 5.4.3.1 of Chap. 5.

In the previous treatments, the objective was to find a seriation structure. Here, it is a classification structure both on the set of column attributes and on the set of row objects. The latter structure provides decorative attribute clusters and jointly clusters of buckles-plates, this without any concern for chronology.

More precisely, we present a couple of classifications, the first on the set of attributes and the second on the set of objects (the plates) in a matrix form. Note that it is the same classification algorithm which is used, with respectively adequate indices, on each of the two sets.

The quality criteria of the clusterings obtained (see Sect. 6.4) lead us to retain the same number of classes for the two partitions, the first on the column attributes and the second on the row objects. In fact, the two partitions that match the best might not have had the same number of classes. It is a particular circumstance which made that the two partitions—on the set of rows and on the set of columns—have the same number of classes; and that, for 5 classes as well as for 7 classes. Thus, 5 classes and 7 classes adequately accommodate the organization of columns and rows. For this, Alg-AL algorithm provided the best results in terms of explained inertia. Therefore, we will detail its results (Figs. 6.14 and 6.15).

In order to reinforce the non-hierarchical clustering of the results obtained, by the methods of the attraction poles, we have applied hierarchical clustering. To this end, average linkage method has been applied to built the ascendant hierarchical classification. In this way, we remain consistent with the criterion used in the non-hierarchical clustering Alg-AL. (see Fig. 6.16).

Partition into 5 clusters				
percentage of inertia explained : 61.47%				
criterion of the adequation quallity : 12.95				
A	B	C	D	E
A8	A7	A1	A17	A19
A10	A9	A12	A13	A14
A21	A3	A6	A23	A11
A20	A22			
A4	A5			
A26	A18			
A16				
A24				
A15				
A25				
A2				

Partition into 5 clusters				
percentage of inertia explained : 70.84%				
criterion of the adequation quallity : 30.70				
1	2	3	4	5
24	1	47	17	2
25	16	54	27	6
32	10	55	51	13
33	38	56	59	12
46	39	57	11	18
49	37	58	48	44
30	40	15	34	19
36	41	23	7	8
3	42	5	50	29
14	43			
31	45			
4				
22				
9				
26				
53				
21				
28				
20				
35				
52				

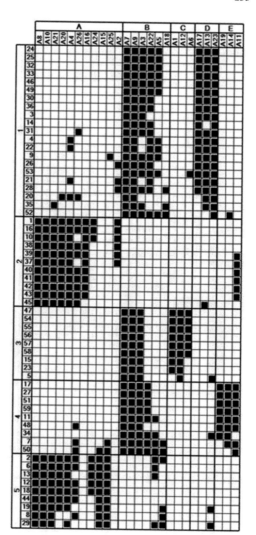

Fig. 6.14 Alg-AL, five clusters for columns and for rows

Partition into 7 clusters						
percentage of inertia explained : 75.74%						
criterion of the adequation quallity : 11.78						
A	B	C	D	E	F	G
A8	A7	A1	A17	A19	A15	A16
A10	A9	A12	A13	A14	A25	A2
A21	A3	A6	A23	A11	A18	
A20	A22					
A4	A5					
A26						
A24						

Partition into 7 clusters						
percentage of inertia explained : 75.33%						
criterion of the adequation quallity : 26.89						
1	2	3	4	5	6	7
24	1	47	17	2	40	28
25	16	54	27	6	41	26
32	10	55	51	13	42	21
33	38	56	59	12	43	
46		57	11	18	39	
49		58	48	44	37	
30		15	34	19	45	
36		23	7	8		
3		5		29		
14						
31						
4						
22						
9						
53						
20						
35						
52						
50						

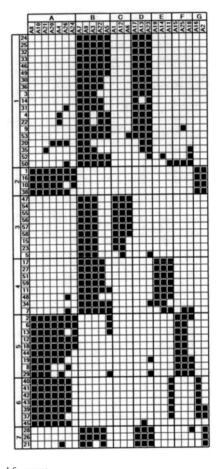

Fig. 6.15 Alg-AL, seven clusters for columns and for rows

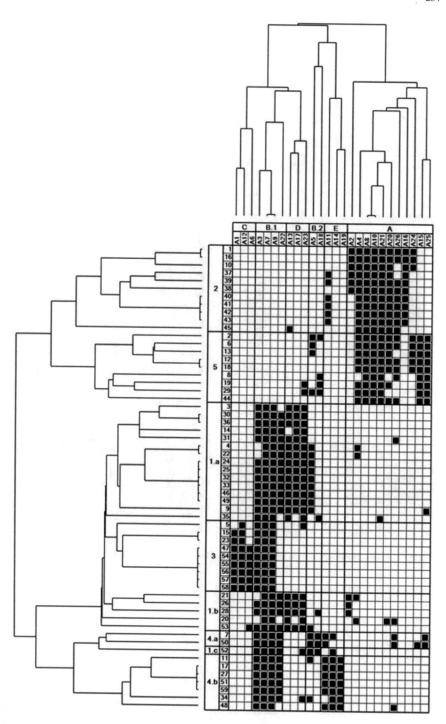

Fig. 6.16 Hierarchical associated clusterings by Average Linkage

Regarding the non-hierarchical classification, we have also considered the following algorithms for the cluster formation: Alg-LLA and Alg-PAL (see Sects. 6.2.4.3 and 6.6.2 and Fig. 6.17).

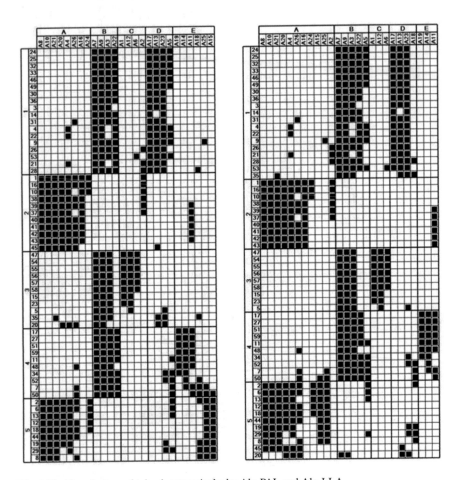

Fig. 6.17 Five clusters, obtained, respectively, by Alg-PAL and Alg-LLA

To finish, we tested the behavior of algorithms coming under the *APMSA* approach
(see Sect. 6.3). And this, in a crossed way (Attributes x Objects) for 5 classes as well
as for 7 classes (see Fig. 6.18). These results can clearly be compared to those obtained
within the framework of the *APMSR* approach (see Figs. 6.14 and 6.15).

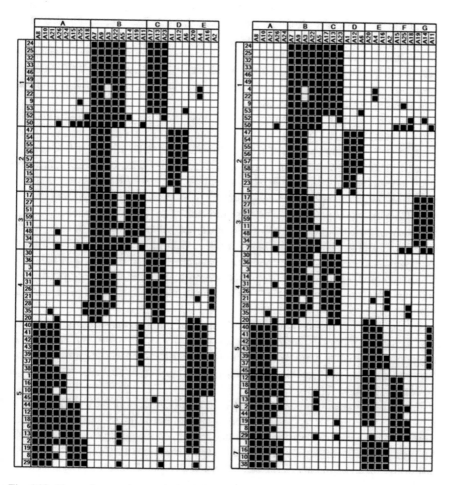

Fig. 6.18 Five and seven clusters obtained by an *APMSA* approach

6.6.5 Some Words to Conclude

In combinatorial data analysis, the mathematical structure sought to synthesize the data is either a partition, or an organized system of partitions, or a total order or a partial order [17]. In the case of partition structures, clustering methods are proposed and in the case of order structures, seriation methods. The orientation of a given method is, very generally, exclusive it is either clustering or seriation.

The literature is very large relative to the first type of structure. Many books have been published in this regard. This is not the case for the second type of structure. Regarding our book, one of the reviewers points out "... on seriation it is a unique book that has a value for publication ... ".

Now, relative to the results we have obtained, it is very interesting to visually perceive the difference between the results obtained in this chapter—oriented towards *classification*—and those of Chaps. 5 and 6, where the orientation is *seriation*.

An approach such as that of Bertin based in its final part, on visual perception is influenced by the two types of structures (see Fig. 4.47). Besides, the first step considered in its strategy includes hierarchical classification of the row set [18].

In the non-hierarchical classification method of attraction poles, each class is drawn by one of the poles. For a given class, the pole that trained it is in a way an extreme representative of the class. In these conditions, it is natural to want to compare our method to that of dynamic clustering—[19] and Chap. 2 of [1]—and this, in the case where each class is represented by its most central point (the one for which the sum of the squares of the distances to the other points of the class is minimum).

This method represents a specific extension of the K-means algorithm. In the absence of any a priori knowledge, the initial choice of attraction centers is made at random and that in relation to their number and their respective positions. And then, it's up to the reallocation and recentering algorithm to converge and to find a clustering structure [7].

Taking into account the fragility of the initialization, it is inevitable that it is necessary to consider a repetition of the algorithm for different initial states. With respect to the sequence of partitions obtained, two types of aggregates are determined. An aggregate of the first type corresponds to an intersection of classes of the different partitions and an aggregate of the second type corresponds to a union of classes of the different partitions.

In our method of attraction poles, thanks to the criterion for evaluating the quality of a partition, we start by determining the right number of poles and the right number of classes. This results in the determination of a partition and therefore, for each of the classes, the possibility of finding its most central element. It then turns out that from these central elements, the reallocation recentring algorithm converges very quickly towards a stable result.

References

1. I.C. Lerman, *Foundations and Methods in Combinatorial and Statistical Data Analysis and Clustering* (Springer Nature, 2016)
2. H.-H. Bock, Clustering methods: a history of k-means algorithms, in *Selected Contributions in Data Analysis and Classification*, ed. by G. Cucumel, P. Brito, P. Bertrand, F. de Carvalho (Springer, 2007), pp. 161–172
3. A.K. Jain, 50 years beyond k-means. Pattern Recognit. Lett. **31**, 651–666 (2010)
4. L. Lebart, J.-P. Fenelon, *Statistique et Informatique Appliquées* (Dunod, 1973)
5. F. Caillez, J.P. Pagès, *Introduction à l'Analyse des Données* (SMASH-BURO, 1976)
6. I.C. Lerman, *Classification et analyse ordinale des données* (Dunod, 1981), http://www.brclasssoc.org.uk/books/index.html
7. I.C. Lerman, Convergence optimale de l'algorithme de "réallocation-recentrage" dans le cas continu le plus simple. R.A.I.R.O. Recherche opérationnelle/Oper. Res. **20**(1), 19–50 (1986)
8. G. Celeux, Étude exhaustive de l'algorithme de réallocation-recentrage dans un cas simple. R.A.I.R.O. Recherche opérationnelle/Oper. Res. **20**(3), 229–243 (1986)
9. H. Leredde, *La méthode des poles d'attraction, La méthode des poles d'agrégation*. Ph.D. thesis, Université de Paris 6, October 1979
10. A.D. Gordon, A review of hierarchical classification. J.R. Stat. Soc. **150**(2), 119–137 (1987)
11. F. Murtagh, P. Contreras, Algorithms for hierarchical clustering: an overview, ii. WIREs Data Min. Knowl. Discov. 1–16 (2017)
12. E.H. Ruspini, Numerical methods for fuzzy clustering. Inf. Sci. **2**, 319–350 (1970)
13. E.H. Ruspini, A new approach to clustering. Inf. Control **15**, 22–32 (1969)
14. L.A. Zadeh, Fuzzy sets. Inf. Control **8**, 338–353 (1965)
15. R.A. Fisher, The use of multiple measurements in taxonomic problems. Ann. Eugen. 179–188 (1936)
16. G.J. Mc Lachlan, *Discriminant Analysis and Statistical Pattern Recognition* (Wiley Interscience, 2004)
17. P. Arabie, L.J. Hubert, Combinatorial data analysis. Annu. Rev. Psychol. **43**, 169–203 (1992)
18. J. Bertin, Traitements graphiques et mathématiques. différence fondamentale et complémentarité. Mathématiques et Sci. Hum. **80**, 60–71 (1980)
19. E. Diday, Une nouvelle méthode de classification automatique et reconnaissance des formes. Revue de Statistique Appliquée **XIX**, 19–34 (1971)

Chapter 7
Conclusion and Developments

7.1 Introduction

We will start by taking a look at what has been done in this book. In the part that stands out, it is an ordinal structure that synthesizes the data. Then, we will indicate developments where the order must be established on all the leaves of a classification tree. These leaves represent elements of a set to be organized. These elements are provided with a similarity index which can be symmetrical or asymmetrical.

As just mentioned, the subject of this book is first of all the seriation problem. Generally, the data can be reduced to a finite set $\mathbf{E}$ provided with a similarity (resp., dissimilarity or distance) index. Formally, a seriation on $\mathbf{E}$ is defined by an ordering (a ranking) on $\mathbf{E}$. Most often, this ordering is specified mathematically by a strict and total order on $\mathbf{E}$. Nevertheless, there are situations in which this total order is relaxed into total preorder (ranking with ties). This ordering has to be built in such a way that, globally, as more as possible, two consecutive elements of $\mathbf{E}$ have the highest similarity (resp., the lowest dissimilarity or distance).

In our development the set $\mathbf{E}$ is defined in the framework of Boolean description of a set $\mathcal{O}$ of objects by a set $\mathcal{A}$ of Boolean attributes. The description representation is an incidence data table comprising zeros and ones, zero for absence of a Boolean attribute on an object and one for presence. Generally, the set of rows is indexed by the object set, while the set of columns, by the attribute set. The set $\mathbf{E}$ may either be the set $\mathcal{A}$ and then, represented by the columns, or the object set $\mathcal{O}$ and then, represented by the rows. This data structure is very basic and very frequent in Data Analysis, in general (see Chap. 3 of [1]) and more particularly in seriation domain. According to [2], the first formal expression of seriation is due to Petrie (1899) [3].

It is in relation to this data structure that we give in Chap. 2 a mathematical definition of the case where a seriation underlies the incidence data table. In practice, a mathematical seriation is exceptionally observed. However, in real cases, a seriation form can express itself statistically with more or less force (strong presence of ones (resp., zeros) in the entries where there must be only ones (resp., zeros)).

I. C. Lerman and H. Leredde, *Seriation in Combinatorial and Statistical Data Analysis*, Advanced Information and Knowledge Processing, https://doi.org/10.1007/978-3-030-92694-6_7

In these conditions, a statistical approach for establishing an association coefficient between column attributes is considered in Chap. 2. The latter leads to a similarity index between the row objects. In addition, we can associate with these similarity indices two dual distance indices, one to compare the column attributes and one to compare the row objects (see (6.18) and (6.19)).

Symmetrical table of association coefficients between column attributes enables an original method of geometrical representation to be built. This method, called Attraction Pole (*AP*) method, is based on the variance analysis of the mutual similarities (associations) between the items to be compared (see Chap. 2).

The first pole of attraction is that of which proximity variance to the other elements is the greatest. The second pole is chosen relatively independent (or distant) from the first pole while having the highest proximity variance to the other elements. And so on ...

The determination of the first two poles makes it possible to draw a planar geometrical representation, revealing a seriation. This is done analogously to geometric data analysis methods such as Principal Component Analysis (*PCA*), Correspondence Analysis (*CA*) or Multidimensional Scaling Analysis (*MDSCAL*). The new method, very simple in its calculation principle is experimentally compared to the latest classical methods (see Chap. 4).

7.2 The Algorithms and Their Respective Logics

7.2.1 Geometric Representations of the Descriptive Attributes and the Objects Described

The first system applied consists of representing by the same method the attribute set and the object set. The planar geometrical representation given by each of the methods *AP*, *PCA*, *CA* or *MDSCAL* provides a layout of the points representing the column attributes or those, the row objects, in a more or less circular or parabolic forms.

In each case, an initial element is determined visually. It corresponds to the origin (or extreme) element of the seriation. From this last, one follows step by step the ordered sequence on the rounded curve associated with the circular or parabolic form mentioned. To make this scan more systematically polar, coordinates are used.

The formal paradigm of this experimental analysis is defined by a σ form with a single moderately chained block, comprising $n = 25$ rows, $p = 16$ columns and $c = 10$ components equal to one per column. Two independent random permutations on columns and rows, resp., are practiced before applying either method. For a given one (*AP*, *PCA*, *CA*, *MDSCAL*), columns and rows are traited separately. It turns out that the ordering results are the same regardless to the method used. The crossing of the two orders (on the columns and on the rows) enables the σ form to be recovered.

The moderately chained σ form (see Sect. 4.4.2.4, Chap. 4) can be considered as a sensitive model for evaluating classical seriation methods.

7.2.2 Row Seriation from Column Seriation

In the algorithm family concerned only seriation (order) on the columns is calculated. And this can be achieved by a geometric method as considered in Sect. 7.2.1 above, or by one of the chaining algorithms as it will expressed in Sect. 7.2.3 below. The order on the row objects leading to a seriation is deduced by means of the combinatorial and statistical algorithm as defined in Sect. 5.2.3 of Chap. 5. The results obtained within the context of the preceding geometric family algorithms for the considered σ form ($n = 25, p = 16, c = 10$) are found. Moreover, compatible results are also obtained for a σ form for which $n = 21, p = 11, c = 11$. However, this is not the case for a weakly chained σ form like the one for which $n = 25, p = 21$ and $c = 5$ (see Sect. 4.4.2.4, Chap. 4).

7.2.3 Algorithms by Successive Chaining

This last example leads us to propose a new and very rich family of combinatorial algorithms. Two parameters govern how to start the seriation (see Sect. 5.3, Chap. 5):

1. By a single element or by an ordered sequence of a few elements;
2. By the manner to aggregate, a new element to the seriation under construction.

The single element starting the seriation could be—as mentioned above—detected visually from the geometric representation provided by one of the methods *AP*, *PCA*, *CA* or *MDSCAL*. Starting with the chosen element, the algorithm proceeds by successive chainings. At each step, a new element is added to the order under construction. This latter element is chosen to be the nearest the last order; for example, the closest according to the similarity index with the preceding extracted element. Here, the seriation (the total order) is assumed to be first established on the column attributes. From there, the order on the row objects is deduced by using the combinatorial and statistical algorithm defined in Sect. 5.2.3 of Chap. 5. The moderately chained σ forms are found. We also find the weakly chained σ form with the parameters $n = 25, p = 21$ and $c = 5$.

Following this interesting result, a new and very rich family of combinatorial chaining algorithms is proposed in Chap. 5. In this family, the principle of successive chaining is retained. Otherwise, an exhaustive set of ways to start the seriation is tested. Finally, as above, we first determine the order on the column attributes and infer the order on the row objects (see Sect. 5.2.3, Chap. 5).

If the initialization is done with a single element, each of the elements (there are p) is tested as a possible start of the seriation. We also consider very original cases for which the initialization is done with an ordered sequence of a given size h where h is of the order of a few units.

For each of the initializations the quality of the reconstruction, in columns and in rows of the seriation form of the data table is measured by various criteria and this gives the most significant seriations.

The developed algorithms are applied to different σ forms. These can consist of one, two or three blocks. In addition and importantly, two real data well known in archaelogy are analyzed through these algorithms.

7.2.4   Attraction Pole Clustering Algorithms

In Chap. 6, we establish a new and very rich family of clustering algorithms. These result from a priori determination of attraction poles. This family is called Attraction Pole Clustering Methods Operating by Successive Aggregations.

A given algorithm of this family produces a descendant but non-hierarchical sequence of classifications. In the latter, each cluster is attracted by an attraction pole. The kth classification (partition, clustering) is into $k + 1$ classes ($k \geq 1$). To assign a new element to one of the clusters under formation, different proximity criteria can be considered. In addition, various quality measures of the partition produced at a given level of the descending process can be examined. These enable us to retain the most relevant partitions and also, to stop the descending formation of the partitions. The treatment of real cases by this clustering algorithmic has been studied.

7.3   Structures and Algorithms in Asymmetrical Hierarchical Clustering

In this section either the notion of symmetric proximity or that of asymmetric one is exploited.

7.3.1   Directed Ascendant Hierarchical Clustering

As just mentioned, the attraction pole method is oriented towards a family of non-hierarchical classification algorithms where , each of the clusters is ordered by proximity with respect to the pole of attraction which attracts it. Very differently, it is a classification algorithm under constraints which leads to a seriation by blocks in the

Marchotorchino method, this on the condition of mutually ordering the blocks by means of a complementary algorithm (see Sect. 3.2.2, Chap. 3).

In [4], methodological proximity between seriation and clustering is underlined. And in this, what is primarily considered is hierarchical clustering.

As expressed in Sect. 4.2.1 of Chap. 3, Bertin used to start by producing a classification tree, obtained by a hierarchical clustering method on the row set of his *matrices ordonnables*. Then, permutations of columns and rows are carried out, using certain manipulations rules and visual perception, this in order to bring out specific patterns.

Seriation on a given set **E**, provided with a proximity coefficient can be obtained by ordering the leaves of a tree corresponding to an ascendant hierarchical clustering on **E**. This order is more directly related to the preordonance associated with the proximity coefficient on **E** (see Sect. 9.3.6 of Chap. 9 of [1]).

More precisely, for a given pair of ordered classes $\{C_1, C_2\}$ merging at a given level of the classification tree, C_1 is placed on the left of C_2 (C_1 before C_2), or C_2 on the left of C_1 (C_2 before C_1), depending on whether the mutual differences in the ranks of the leaves respect the preordonance more.

This underlies that basic principle adopted in [5]. Brossier considers different types of exogenous variables (parameters) to respect on the set hierarchically clustered. For each of them and independently, the leaf order has to follow as well as possible the variable behavior. One of the considered variables is a distance on the clustered set. However, the considered case above of the preordonance associated with this distance is not studied (see Chap. 9 of [1]).

The last analysis by directed hierarchical clustering uses a matrix of symmetrical dissimilarity between the unit elements. It must clearly be distinguished from the case where the dissimilarity matrix is D is asymmetrical. In the latter case, D can be decomposed additively into two parts D_a and D_s, where D_a is the antisymmetric part and where D_s is the symmetric one. More precisely, T denoting *Transpose*, we have

$$D_a = \frac{1}{2}(D - D^T) \text{ and } D_s = \frac{1}{2}(D + D^T) \tag{7.1}$$

The preference, exchange flows, migration or confusion rate between stimuli are of this type of matrices.

In [6], the ascendant hierarchical clustering is established on the basis of the symmetrical matrix D_s, where a given value $D_s(o_i, o_j)$ of elements (objects) ($1 \leq i, j \leq n$) is defined by the sum of the exchanges from o_i to o_j and from o_j to o_i.

Now, regarding to D_a, it is approximated by an antisymmetric matrix of the form

$$E = \{E_{ij} = e_i - e_j | 1 \leq i, j \leq n\} \tag{7.2}$$

where

$$e_i = D_a(i.) = \sum \{D_a(i, j) | 1 \leq j \leq n\}$$

e_i defines the elongation of the ith branch leading to the leaf representing the object o_i, $1 \leq i \leq n$.

Then, the leaves of the tree are arranged as much as possible without crossing branches and this, according to the decreasing values of the e_i, $1 \leq i \leq n$.

In the model proposed in [6], the hierarchical clustering is only based on the symmetric matrix D_s. However, both D_s and D_a are employed in [7]. Let us point out here that a general and detailed view of reactualization formulas (see Chap. 10 of [1]) in the case of an asymmetric ascendant hierarchical clustering is given in [8, 9].

7.3.2 Oriented Hierarchical Clustering

We consider as above an incidence data table composed of zeros and ones and resulting from the Boolean description of a set $\mathcal{O}$ of objects by a set $\mathcal{A}$ of Boolean attributes. In this context, the symmetrical notion of an association coefficient between Boolean attributes has been extended to an asymmetrical version. Such a coefficient is called *implication* index. Let us denote the square table of these indices between the Boolean attributes as follows:

$$\{\mathcal{I}(a, b)|(a, b) \in \mathcal{A} \times \mathcal{A}\} \tag{7.3}$$

where $\mathcal{I}(a, b)$ is a measure of the intensity of the implication of a towards b ($a \longrightarrow b$). Intuitively, this measure expresses how much a *TRUE* value of a implies a *TRUE* value of b. There may be different versions of the implication index. All of them are *asymmetrical*. By difference, note that in seriation problem, the similarity coefficient between two column attributes is essentially symmetrical.

A specific variant of the implicative index which has been particularly worked on derives from the symmetrical coefficient (2.9) of Sect. 2.3.2 of Chap. 2, [10, 11]. Let us imagine that we can associate a distance index with the implication index in the same way as we associated the distance index D_α with the similarity one S in the symmetric case (see (6.36) and (6.38) in Sect. 6.2.1 of Chap. 6). In these conditions, it is possible to envisage the same analysis as that considered in Sect. 7.3.1.

Intuitively, the implication holds mathematically if $\mathcal{O}(a)$ is included in $\mathcal{O}(b)$, where $\mathcal{O}(a)$ (resp., $\mathcal{O}(b)$) is the subset of objects where a (resp., b) is *TRUE*. Such an implication must be evaluated all the more strongly by the index $\mathcal{I}(a, b)$ as $\mathcal{O}(a)$ is a large subset of $\mathcal{O}(b)$ and as the cardinal of $\mathcal{O}(b)$ is small. In fact, more often than not, $\mathcal{I}(a, b)$ has to deal with situations in which $\mathcal{O}(a)$ is strongly but not strictly included in $\mathcal{O}(b)$.

A first coefficient has been proposed by J. Loevinger [12], but it does not satisfy the mathematical inclusion condition just expressed. Besides, in real cases where mathematical inclusion occurs, a given value of this index cannot be interpreted, especially when it comes to compare mutually several pairs of behaviors.

R. Gras [13] got the idea of adapting our method in constructing a symmetrical association coefficient of the likelihood of the link between Boolean attributes [14] and Chap. 5 of [1] to the case of an asymmetric (oriented) association. In this, for a given ordered pair (a, b) of Boolean attributes, $\mathcal{I}(a, b)$ refers to a probability scale defined with respect to an independence hypothesis of no link.

The hypothesis considered in [13] refers to a binomial model. The construction of this approach has been completely resumed [14]. We were led to sustitute the Poisson model for the binomial one. The new model and the resulting coefficient have been adopted by R. Gras and P. Kuntz [15].

From such an implicative coefficient giving rise to a matrix such (7.3), construction of an oriented tree on the attribute set is proposed in [10]. This tree is asymmetrical and obtained by directed hierarchical clustering algorithm. In the latter the link between two clusters is oriented from left to right. More precisely, if (C_1, C_2) is an ordered pair of classes (C_1 at the left of C_2) such that a node connects C_1 to C_2, then there is a global implicative tendency from the elements of C_1 to those of C_2.

The mathematical analysis of this structure is provided in [11, 16] and reported in Sect. 1.4.5 of Chap. 1 of [1]. Now, different similarity implicative indices between (C_1 and C_2) (from (C_1 to C_2)) can be proposed; more particularly, the oriented index of the likelihood of the maximal link [11, 15]. Such an index can be denoted by $\mathcal{I}(C_1, C_2)$.

7.3.3 Hierarchical Clustering of Successive Intervals

In [17] construction of ascendant hierarchical aggregations on a totally ordered set **E**, is proposed. For starting the aggregation process, this order is chosen arbitrarily. It evolves as the algorithm progresses. **E** is assumed to be provided by a symmetrical dissimilarity index between disjoint subsets of **E**.

In the classical ascendant hierarchical clustering, the hierarchy of the levels (also called nodes) determines a family of subsets of **E** such that two of them are either disjoint or one included in the other.

The structure concerned here—called *pyramidal* hierarchy—gives rise to a family of intervals such that, two of them are either disjoint, have a non-empty interval intersection belonging to the family, or one included in the other.

As just mentioned, the initial arbitrary order adapts step by step, so as to avoid crossing between branches of the hierarchical tree.

In this ordinal construction, the dissimilarity between two intervals is exactly the same as that between the two respective subsets included in these intervals. This dissimilarity is symmetrical and does not take into account the order of the elements.

In the case of classical binary hierarchical clustering the dissimilarity index between two classes G and H ($G \subset \mathbf{E}$, $H \subset \mathbf{E}$ and $G \cap H = \emptyset$) does not consider the respective hierarchies of subsets leading to G and H.

In the pyramidal case the dissimilarity index between two intervals I and J involves the respective interval pyramids leading to I and J, resp.

In the classic hierarchical clustering a given cluster G can be related to a maximum one cluster H on its right or on its left, exclusively. In a pyramid layout, a given interval of the concerned family, can have two junctions, one on its left and one on its right.

In the pyramid case, the dissimilarity between two constituted intervals (two levels) I and J, is defined by the minimal dissimilarity between all the ordered pairs of nodes, without double junctions, where the first (resp., the second) participates in the formation of I (resp., J). The result obtained is generally difficult to control.

Now, consider in a pyramidal hierarchical structure the set L of all of its levels (nodes) linked both to the left and the right to its adjacent neighbors. By deleting for each of the L levels, one of their two junctions, we obtain a classical binary hierarchical clustering. In this regard, it may be interesting to remove—each time—that junction associated with the greatest dissimilarity between levels.

The pyramidal representation is presented in [17] as a generalization of classification tree representation associated with hierarchical clustering. In fact, it is a very particular representation.

Considering the arbitrariness of the initial order and also, the symmetrical nature of the dissimilarity, it is difficult to realize the capacity of the pyramid technique to turn out seriations, and this, more particularly, compared with other methods. Examples are given in Sect. 6 of [17].

7.3.4   Some Additional Interactions Between Seriation and Hierarchical Clustering

Let us return to the seriation problem in which the similarity index between two column attributes (resp., row objects) is symmetric. In these conditions, suppose given a σ form as defined mathematically in Chap. 2 and made up of more than a single block (see Fig. 2.10 of Chap. 2 and Figs. 4.2 of Chap. 4). Now, consider applying the well-known single linkage ascendant hierarchical clustering (see Sects. 10.2.2 and 10.3.4 of Chap. 10 of [1]) to the set of column attributes. An adaptation of this algorithm makes it possible to bring out the σ form in all of its components.

More precisely, by applying the multiple aggregation version of this algorithm (see Sect. 10.6.3.2 of Chap. 10 of [1]), the subdivision into different blocks of the σ form is obtained at the first level of the classification tree.

Now, consider the ordered sequence of binary aggregations with respect to any one of the blocks, regardless of the latter. The first binary aggregation is made between two consecutive column attributes, those encountered first by the algorithm. At a given non-final step, an ordered sub-block of one of the one of the blocks is constituted.

Indeed, at a given level, a column attribute is sought so as to be the nearest of the sub-block being formed and this, according to the single linkage method. There can be two solutions corresponding to the two ends of the sub-block in construction. One of them is chosen and placed adjacent to the end of which its the closest.

Mostly (see above in this section), we try to induce a seriation from the ascendant formation of a hierarchical binary classification. In the case of Murtagh development for analyzing "very high-dimensional data, and, quite often, very big data sets and massive data volume", the matter consists of inducing hierarchical clustering from seriation [18, 19].

A final point. In the developed research in this book, seriation is formalized by a total order on the set concerned (a set of rituals or, dually, a set of tombs in the case of archaeology). However, often, the expert knows some of the inequalities of the order to seek for seriation. Thus, in relation to archaeology, the archaeologist can express that such a ritual r precedes such other ritual r', or, such a tomb t is prior to such another tomb t'.

In these conditions, we are faced with the problem of finding a total order under constraints, optimizing globally the mutual similarities between consecutive elements. This is a fairly important research topic.

References

1. I.C. Lerman, *Foundations and Methods in Combinatorial and Statistical Data Analysis and Clustering* (Springer, 2016)
2. D.G. Kendall, Seriation from abundance matrices, in *Mathematics in Archaeological and Historical Sciences*, ed. by D.G. Kendall F.R. Hodson, P. Tautu (Chicago, Aldine-Atherton, 1971), pp. 214–252
3. W.M.F. Petrie, Sequences in prehistoric remains. J. Anthropol. Inst. **29**, 295–301 (1899)
4. I. Liiv, Seriation and matrix reordering methods: an overview. Wiley InterScience (www.interscience.wiley.com), (DOI: https://doi.org/10.1002), 70–91, Jan. 2010
5. G. Brossier, Algorithmes d'ordonnancement des hiérarchies binaires et propriétés. Revue de Statistique Appliquée **32**, 65–79 (1984)
6. G. Brossier, Classification hiérarchique à partir de matrices carrées non symétriques. Statistique et Analyse des Données **7**, 22–40 (1982)
7. D. Vicari, Classification of asymmetric proximity data. J. Classif. **13**, 386–420 (2014)
8. H. Yadohisa, Formulation of asymmetric agglomerative hierarchical clustering and graphical representation of its result. Bull. Comput. Stat. Jpn. **15**, 309–316 (2002)
9. A. Takeuchi, T. Saito, H. Yadohisa, Asymmetric agglomerative hierarchical clustering algorithms and their evaluations. J. Classif. **24**, 123–143 (2007)
10. R. Gras, A. Larher, L'implication statistique, une nouvelle méthode d'analyse des données. Mathématiques et Sciences Humaines **120**, 5–31 (1993)
11. I.C. Lerman, Analyse logique combinatoire et statistique de de la construction d'une hiérarchie binaire implicative ; niveaux et noeuds significatifs. Mathématiques et Sciences Humaines/Mathematics and Social Sciences **184**, 47–103
12. J. Loevinger, *A Systematic Approach to the Construction and Evaluation of Tests of Ability.* Psychological Monographs, vol. 61, number 4 (1947)
13. R. Gras, *Contribution à l'étude expérimentale et à l'analyse de certaines acquisitions cognitives et de certains objectifs didactiques en mathématiques, doctorat d'état.* Ph.D. thesis, Université de Rennes 1, Oct. 1979
14. I.C. Lerman, R. Gras, H. Rostam, élaboration et évaluation d'un indice d'implication pour des données binaires i et ii. Mathématiques et Sciences Humaines **74** and **75**, 5–35 of 74 and 5–47 of 75 (1981)

15. R. Gras, P. Kuntz, An overview of the statistical implicative analysis development, in *Statistical Implicative Analysis*, ed. by F. Guillet, R. Gras, E. Suzuki, F. Spagnolo (Springer, 2008), pp. 11–40

16. I.C. Lerman, P. Kuntz, Directed binary hierarchies and directed ultrametrics. J. Classif. **28**, 272–296 (2011)

17. P. Bertrand, E. Diday, Une généralisation des arbres hiérarchiques : les représentations pyramidales. Revue de Statistique Appliquée **38**(3), 53–78 (1990)

18. F. Murtagh, P. Contreras, Algorithms for hierarchical clustering: an overview, ii. *WIREs Data Mining Knowl Discov 2017* (2017), pp. 1–16

19. F. Murtagh, *Data Science Foundation; Geometry and Topology of Complex Hierarchic Systems and Big Data Analytics* (Chapman and Hall, 2018)

Index

A

Aggregate, 5, 216, 251, 260, 265

Aggregation, 189, 229, 241, 269, 270

Algorithm, 3, 5, 6, 28, 41, 43, 52, 56–61, 64, 65, 70, 72–74, 90, 95, 101, 103–108, 116, 118–120, 122, 124, 125, 127, 130–132, 134–138, 141, 143, 146, 148, 149, 153, 160–163, 168, 170, 172, 175, 179, 182, 184, 187, 188, 191–200, 202, 204, 205, 210, 211, 215, 217, 222, 226, 228–231, 236–241, 250, 251, 254, 258–260, 265–267, 269, 270

Algorithmic, 2–5, 7, 11, 60, 65, 66, 70, 74, 86, 105, 107, 112, 119, 131, 176, 192, 229, 230, 266

Analysis, 1, 4–8, 12, 14, 17, 18, 20, 34, 39, 43, 47, 55, 58, 60, 61, 64, 66, 72, 81, 90, 101–103, 105, 106, 108, 111, 115, 117, 120, 122, 132, 140, 147, 162, 163, 185, 224, 232, 238–240, 264, 267–269

Archaeological, 147, 161

Archaeologist, 2, 12, 147, 148, 160

Archeology, 6, 271

Arrangement, 5, 45–47, 60, 72, 73, 106, 116, 176, 177, 186–192, 211

Assignment criteria, 223, 231

Association, 8, 12, 71, 79, 80, 106, 176, 183–186, 189, 216, 230, 236, 239, 264, 269

Association coefficient, 6, 7, 12, 14, 17–20, 43, 45–47, 55, 63, 66, 78–80, 84, 86, 88, 99, 100, 114, 115, 176, 183, 218, 236, 264, 268, 269

Asymmetrical, 1, 2, 263, 267–269

Asymmetrical hierarchical clustering, 266

Attraction, 6, 25, 47, 48, 50, 55, 56, 60, 66, 86, 102, 107, 108, 112, 115, 116, 119, 126, 128, 176, 188, 215–217, 220, 222, 223, 229–231, 236–238, 240, 241, 254, 260, 264, 266

Attraction poles method, 4, 6, 39, 47, 60, 102, 128, 134, 140, 145, 154, 168, 215, 229, 237, 239, 260, 264, 266

Attribute, 1–8, 11–15, 17–20, 24, 25, 32, 39, 42, 43, 45, 47–50, 55–60, 62–71, 73, 74, 78, 81, 84, 86–88, 99–102, 104–108, 113–120, 128–130, 132, 138, 146, 159, 161, 163, 175–179, 183–185, 191, 196, 204, 216–220, 223, 224, 227, 229, 234, 235, 237, 239, 240, 254, 259, 263–265, 268–270

Axe, 47, 49, 50, 55, 60, 102, 115–117, 126

B

Bi-classification, 6

Bi-clustering, 6

Binary, 55, 59, 66–68, 70, 78–80, 84, 86, 88, 179, 181, 184, 269–271

Binomial, 19, 75, 226, 228, 269

Block seriation, 6, 51, 58, 59, 63–65

Bond energy algorithm, 108

Boolean, 3, 5–8, 11, 12, 14, 17–19, 45, 47, 55, 56, 58–60, 62, 63, 66–72, 74, 99, 100, 102–106, 108, 113, 115, 159, 175, 176, 218, 220, 225, 239, 240, 263, 268, 269

Buckles-plates, 111, 112, 120, 147, 203, 205, 206, 239, 240, 254

Printed in the United States
by Baker & Taylor Publisher Services